The British Police and Home Food Production in the Great War

Mary Fraser

The British Police and Home Food Production in the Great War

Police as Ploughmen, 1917–1918

palgrave
macmillan

Mary Fraser
School of Social and Political Sciences
University of Glasgow
Glasgow, UK

ISBN 978-3-031-58745-0 ISBN 978-3-031-58743-6 (eBook)
https://doi.org/10.1007/978-3-031-58743-6

This Palgrave Macmillan imprint is published by the registered company Springer Nature Switzerland AG
The registered company address is: Gewerbestrasse 11, 6330 Cham, Switzerland

To Peter and Jane who have always been so supportive

Preface

Interest in the history of the British Police is developing apace. Embedded in the social and political culture of the time, police work is often surprising and shows their compassionate nature as well as their tough character. They are unique individuals, and as a service they illustrate how to manage the population in times of crisis, calming frightened and distressed people without adding to the trauma of the situation. The police are autonomous from government, responding to national emergencies by understanding the public mood. This explains why one of the best-known phrases about their role is: the police are the people, and the people are the police.

Glasgow, UK Mary Fraser

Acknowledgements

The author is most grateful to the Police History Society for the small grant which allowed travel to visit archives around Britain; the extent of this work would not have been possible without it.

Many archivists helped to access the material for this book: Alyson Pigott, Tyne and Wear Archives; Information Rights, Essex Police; Jacquie Crosby, Lancashire Archives; Brian Phillips and Philip Bye, East Sussex Record Office; Berni May, West Midlands Police Museum; Tom Townsend, Norfolk Record Office; Hannah R. Bell, Live Borders, Hawick; Caroline Picco, Cheshire Archives and Local Studies; Ashleigh Black, Aberdeenshire Archives; Roger Hull, Liverpool Record Office; Dr Helen Wicker, Kent Libraries, Registration and Archives; Jane Bass, Essex Record Office; Linda Bankier, Berwick Record Office; Kathy Hobkirk, Live Borders; Richard Callaghan, Royal Military Police Museum, Southwick; Jeff Walklate and Jon Mein, Hertfordshire War Society; Debbie Burstow, Lynne Burton and Gary Moyle, Hertfordshire County Council Heritage Services; Steve Wilson, Preston Archives; Alice Millard, West Sussex Record Office; Alasdair Malcolm and Catherine Glover and Professor Richard Hoyle of the British Agricultural History Society; and Dr Stefan Goebel.

During preparation of this book many people have offered an opportunity to present the initial findings and I have gained valuable feedback from the interactions, they include The Social History Society conference 2019; The Royal Philosophical Society of Glasgow 219th lecture series in October 2020; the British Agricultural History Society Spring seminar series April 2021; the British Crime Historians symposium 2021, University of Leeds; Policing and Society conference, University of Akureyri, Iceland, 6th October 2021; the Police History Society conference November 2021 and the Institute of Historical Research seminar series War History and Society 31st May 2023.

Contents

List of Figures

List of Tables

in Chapter 6 with farmers who refused to plough their land to increase crop production.

This book demonstrates how the rise in importance of home food production to combat the fear of national food insecurity gave rise to the establishment of state control of agriculture, enabling the police to work closely with local agricultural committees by providing practical help to farmers. Police presence and their help on farms were welcomed, as they showed competence in the work, therefore allowing close surveillance of farming to carry out the aims of the state. But as well as their close work with farmers, the police were also ready to quell potential or actual rioting, if the population's fears of starvation ran high due to threats to themselves, their families and communities, confronting disorder directly. This example of police as ploughmen extends police work far beyond law enforcement [2, 3] into the protection and fostering of social order by working alongside those providing help for the population in distress; the police become an institution of administration of the population, rather than of the exercise of law [4].

The police role in helping the population in distress while providing a service is not only seen during the First World War, it is an enduring role in British policing, an historical continuity [5], it helps to explain the present, most recently during the pandemic when the police were again released, this time as ambulance drivers, although the nature of their help, the strategies to provide it and the demands on them changed over time.[1] These examples bring policing by consent closer to the people, while enhancing British police legitimacy and effectiveness [6, 7], allowing early detection of points of conflict and potential unrest. The continuity of lending police help to other services in times of crisis informs today's police that this role remains a valid use of resources, despite the test of time.

This book shows how the police, by working closely with people affected by a national crisis, provide a valuable service, in this case to farmers, to prevent or minimise unrest where rioting due to food deprivation was already seen in reports from other combatant nations. I use the definition of crisis as a time when the population looked to government to solve the shortages of basic foodstuffs which appeared to be escalating out of control with no end in sight, threatening to lead to

[1] This will be developed further at the end of Chapter 9.

population starvation [8]. This kind of police help is best shown in a snapshot or micro-historical social history [9] at a turbulent time. We see how policemen[2] worked within the dominant discourse of patriotism to help not only their local population in crisis within the boundaries of their area, but also more distant populations in some instances. The police, as first responders, acted before most other groups and made a prior assessment of their skills to ensure appropriate help was given melding into the service, so that those who they worked alongside welcomed their intervention.

Today the repeating theme of helping the population in distress is learned early in modern police training, when recruits progressively appreciate that their role is more about protecting citizens and society and maintaining social order, than about catching criminals [10, 11]. Indeed, the College of Policing in 2015 found that calls on police help were increasingly about incidents that involved public safety and welfare [12].

This book provides the detail of the historic response by the police to the national crisis of food shortages with the widespread fear of population starvation in 1917, mid-way through the war in the year characterised by war weariness. By 1917 the war had dragged on for more than three years with huge loss of life in battle and little progress being made in struggles to contain or win ground against the enemy. The fierce battles in Europe were based on killing the enemy by close combat in trench warfare, with atrocious land conditions, such as in the latest battle, the Somme (July to November 1916), one of the costliest battles of the First World War in which Britain suffered 57,420 casualties in the first day alone, the largest loss of life to date by the British Army in a single day. It was the first battle fought by many of the men recruited into the new army in 1914,[3] therefore would have had many repercussions at home. Reports from the front, along with pacifist protests and conscientious objectors at home insisting the war was unnecessary, advocating immediate peace negotiations, contributed to increasing war weariness in a population fatigued by war and wondering if, and when, it would end, while mourning their dead in battle [13]. Consequently, the threat of food insecurity, mainly due to blockading of food imports, added to the

[2] At this time the police was an all-male occupation.

[3] https://www.iwm.org.uk/history/key-facts-about-the-battle-of-the-somme (accessed 2/3/2024).

population's distress and became a national emergency with the urgent need to increase home food production reaching crisis point in March 1917, as Offer says "agrarian resources decided the war. So not only a war of steel and gold, but a war of bread and potatoes" [14]. This reflects Lloyd George's statement in the third volume of his memoirs "the food question ultimately decided the issue of this war" [15]. Without rapid intervention, food riots would have increased, as they did in other combatant nations, such as Germany, Austria-Hungary, Russia and France [16]. With a speedy response in Britain in Spring 1917, policemen were allocated temporarily to farms to help farmers increase crop production in the national interest, some forces continued this help until the end of the war, often at intervals of maximum need, such as Spring ploughing and planting and autumn harvesting.

The initial escalating emergency was prominently reported daily in national and local newspapers in February, March and April 1917 increasing the alarm in the population by the urgency of reports that there was only six weeks remaining to plough and plant for an autumn harvest, if national food shortages were to be avoided, as shown in Fig. 1.1.

National government-led mobilisation was quickly developed to help agriculture: the army released soldiers from the Home Front and the Department of National Service requested local authorities and employers to temporarily release their staff with agricultural skills to help farmers. Policemen were an ideal group to target, as in the second decade of the twentieth century the backgrounds and previous occupations of many were from agriculture, and it was generally known that many policemen possessed the required skills:

> The Chief Constable of Glasgow has been requested by the Special Committee on Agricultural Produce to give a return of the number of Policemen in the City who have had agricultural experience. …
>
> MR. WHEATLEY (a Glasgow councillor) said the Police was partly a Government Force, and as the Government were appealing to other employers to release their men, surely the Corporation might approach the government and point out that there must be thousands of policemen all over the country who might be released …. [18]

This quotation from a meeting of Glasgow Corporation during discussion of how the local authority could help to increase home food production shows the widely held view that the police possessed the skills

Fig. 1.1 National service advertisement [17]

and abilities to provide appropriate help to farmers. Indeed, it had been generally known in rural communities since at least the mid-1800s that men who desired social advancement entered the police force [19].

Recruitment patterns into the British police during mid to late nineteenth century confirm the targeting of men between 20 and 27 years of age from rural communities due to their strong physique giving them a commanding appearance, conspicuous on the streets as a symbol of

recruited largely from Inverness or from the bordering counties of Ross, Moray, Sutherland and Caithness. The county of Caithness, Orkney and Shetland's recruits were even more local, born within the district itself. The picture differed in Glasgow where recruitment was largely from the highlands, and the north-east with largely farming communities, and from Ireland, making up around three-quarters of Glasgow City Police between 1900 and 05, although local recruitment increased from 1909 [31, 32]. The County Forces in England were less than half of those who released policemen into agriculture in 1917 and 1918 as we see in Chapter 6. But their role in both England and Scotland assigned by government to distribute and collect censuses by hand, as in the 1916 census of farm labour and 1917 survey of horses, (see in Chapter 3, p. 48–49 and Chapter 4, p. 89) increased their acquaintance with individual farmers; and.

3. The City and Borough Police Forces numbered 128 in England and Wales with a strength of 19,000; they covered the towns and cities. In Scotland there were 29 City and Burgh police forces. Between 1889 and 1920 only towns and cities with populations above 10,000 were permitted to have a police force, smaller forces were amalgamated into County constabularies in the Local Government Act of 1888 [33]. Each City or Borough (Burgh in Scotland) Force was managed by the Town Council's Watch Committee responsible for appointing the Chief Constable and policemen. Their independent status from the Home Office or Secretary of State was jealously guarded by local authorities. Recruits to the Birmingham force were largely born outside the city, while in other more remote areas such as Ipswich, more than 50% were born locally. Birkenhead was notable in recruiting nearly a half who were born outside the bordering counties with over a quarter born in Ireland. In Scotland the combined establishment of County and Burgh police was 5,953, with Glasgow containing the largest establishment of 1,996 [34, 35]. These recruitment patterns show the likelihood of policemen allocated to agriculture in 1917/18 being known to farmers and is discussed further in Chapters 6–8.

On the outbreak of war in 1914 police manpower was mostly of military age and contained military reservists who were called up immediately. Others volunteered in the national fervour of recruitment during the first year of war. Initially replacements were employed temporarily for the duration of the war, but they became largely ostracised by their colleagues as war progressed claiming they were using the police as a way of evading conscription, so were deeply unpatriotic. We see in the example of Birmingham in Chapter 8, how they were the first to be released into the army in early 1917. But the service was not allowed to recruit to fill vacant posts from 1916, due to the increasing need for national manpower planning and competition with the army for fit young men of military age [36].

The early withdrawal of policemen into the military had depleted forces by around one in five in the provinces and approximately a quarter from the Metropolitan Police by the end of 1915. In addition, the police lent a few constables to the army as drill instructors [37]. The diminution of regular policemen raised alarm in the service, with protests and claims by Chief Constables that their force had lent as many men as could be spared, by 1917 the loss of men had inevitably increased:

> SUPT. MEARS said the full strength of the Force in his division was 54. He had lost 20 men since the outbreak of war, and was still 16 short. The Chief Constable was tied down in this matter by the War Office and the Home Office. [38]

This example shows a loss of 37% of the Chertsey division. The numbers recruited into the army from larger forces were in similar proportions; Glasgow had released 745, 37.2% of its force by mid-June 1917 [39] and smaller borough forces such as Reigate lost 32% recruited into the army[40].[4] As well as these losses, periodic combing out of policemen continued from the implementation of conscription in the Military Service Act 1916, so it is initially surprising that any further policemen could be spared for release into agriculture in Spring 1917.

Despite these huge losses and under pressure from the discourse of patriotism, the police service did not collapse with the weight of recruitment into the armed services, nor was it prioritised and protected as an essential service, with Chief Constables and local authority police committees left to decide how to maintain their numbers; how individual forces did this was often contentious and will be shown in Chapters 6–8. As with most war services, attempts at substitution were made with

[4] The contact for Robert Bartlett is his email address, given here with his permission 938at938@gmail.com.

several measures adopted, for example, the Police (Emergency Provisions) Act 1915 [41] suspended the right to retire except with the Chief Constable's permission; those already retired and receiving a pension were re-employed; the number of part-time Special Constables was dramatically increased, and they progressively performed ordinary police duties on the beat. All these measures contributed to making the police service increasingly composed of older men with part-time volunteers. The weekly rest day granted in 1910 was reduced or abandoned as was annual leave in many forces, with some angry protests [42].

Furthermore, police duties did not continue as before, the war required additional tasks, such as enforcing blackouts, preventing photography, sketching or note-taking in sensitive areas, and carrying out various censuses, which the remaining policemen and substitutes had to perform with no direction from government on priorities [43]. In addition, the police responded to periodic concerns, such as in 1914/15 the request by the Home Office to check the character of all women receiving the war separation allowance to ensure they were of a sufficient moral character to receive it, in 1916 to check certificates of exemption from conscription under the Military Service Act 1916 and in 1917 requests to look out for organised agitation aimed to prevent munitions work [44–47]. Bearing in mind these drastic changes in the composition of police forces and the additional duties, Chapters 6–8 show how the police overcame the dilemma of the release of more men away from front-line police duties into agriculture and Chapter 9 gives an indication of why it could be in their best interests.

The dominant national discourse of patriotism was very evident in the work of the police during the First World War, as they represented the rule of law. As a key institution, they mediated the discourse and its effects on the population. Patriotism stood for the love of and loyalty to one's country, a principal that all those living in the country were expected to share, it says that the country as a whole is pretty good, and we don't want to change it, our country is worth defending [48].

The pressure of wartime heightened and intensified patriotism, involving the evaluation of others living in the country. Individuals and organisations were evaluated as either the in-group who supported patriotism by opposing the enemy, or the out-group who were either conscientious objectors, pacifists or opposed aspects of the war such as the Anti-Conscription League. Out-group members were evaluated by individual's commitment to the country's good and their contribution to it. This allowed identification of the good or bad individual or organisation [49]. At an institutional level the police was a symbol of the patriotic state, encouraging and enforcing solidarity to the national cause of fighting for the country [50]. This involved carrying out the national agenda, such as checking individuals for possession of a Certificate of Registration, required under the National Registration Act 1915 [51] and rounding up men who evaded conscription into the army following the first Military Service Act, labelled "shirkers" [52], with the approval of the Local Government Board [53]. The national institutions by working together ensured that the discourse of patriotism circulated in the population trying to prevent anyone escaping its impact. At an individual level, each policeman was expected to adhere to the principal of love of one's country and to do everything he could to help to win the war, as part of his personal identity [54]. This book shows the struggle for the police and the pressure on the service to adhere to these principles of patriotism which increased as war progressed, and how individual policemen of military age struggled to commit to its demands to go to war, as until late 1917 the nation recognised that the soldier had a higher priority than the citizen. From late 1915 many men of military age did not want to go to war, as seen in the dismal failure of the Derby Scheme,[5] the last-ditch attempt by the Liberal government to avoid compulsion, to recruit sufficient men to meet the army's needs, as many found ways to resist joining up; policemen were no exception. As the police was largely made up of men of military age, this posed a significant difficulty for the individual policeman and for the organisation of policing. The British police reliance on public consent meant maintaining public approval by their patriotic actions was vital, they should show unity with the population and a sense of shared identity and fate in a good outcome from the war.

This book sees the police reluctantly adhering to public pressure by progressively not deploying men of military age on the beat where they were highly visible and shown to be unpatriotic by not supporting the aims of the army to recruit fit young men. At a time when most police work was on the beat on the streets in the towns and cities, they should

[5] The Derby Scheme secured pledges from all men aged 18–40 to join the army when called, by either joining immediately (around 11% did so) or deferring their army service and were then called up in sequence, the youngest single men first. However, without compulsion, only two-thirds of single men and half of the married men signed up, the remainder avoided recruitment (Fraser 2019 p. 140).

days it encouraged the moral values of stability, self-improvement, thrift and sobriety and had a clear view of the policeman as respectable, self-disciplined and self-taught, which was close to the principles advocated by police officials. It spoke the language of self-help, fair treatment and improved working conditions, which contributed to police work being seen as a respectable career with promotion prospects. During the First World War, police officers were vulnerable financially due to poor pay, restricted in their ability to organise themselves to press for better conditions and to be recognised as having specialist skills, their weekly journal therefore performed a valuable role in providing a voice [61]. Figure 1.2 shows the typical front cover of *The Police Review and Parade Gossip* during its early years.

Widely read and influential journals such as *The Police Review* are important sources for historical research because in order to continue to attract and keep their readers, any newspaper or journal must reflect readers' views and, with an occupational journal, the views of the organisation in which they work. We therefore have a sense of the legitimate issues of the day for the police and their struggles through reading this journal. However, *The Police Review,* like most journals, was not impartial; its influence was to show readers how to behave and work as the good policeman and citizen and to uphold the moral values of the day.

Police as Ploughmen, after initial identification as column headings in *The Police Review,* was followed up by searches in local and national newspapers, local and national archives, parliamentary debates, well-reviewed books and papers and other public and private sources such as police museums. Searches were guided first by the wish to establish the authenticity of reports in *The Police Review,* this was by comparison with local and national newspaper reports and archival material and when established, to ask why the actions occurred and what was the social and political background that made them possible. Searches were conducted until saturation was achieved [62].[7]

Chapters 2 to 5 give the national and local conditions that made the release of policemen into agriculture possible. Chapter 2 develops the social and political crisis that led to the initial food shortages from March 1917. It sets the scene by giving the recent history of agriculture up to the outbreak of the Great War in August 1914, to show how and why

[7] Saturation is defined as "We stop gathering more cases when we repeatedly find the same answers."

WAR PRICE 2d.

POLICE THE REVIEW
& PARADE GOSSIP:
ORGAN OF THE BRITISH CONSTABULARY

THE "SAMSON" LEVER
FULL HUNTING DOUBLE CASES.
SOLID SILVER CASES
85/- SPECIAL OFFER TO POLICEMEN

POLICE!! POLICE!!!

SEVEN DAYS' FREE TRIAL

FREE!
A FINE CATALOGUE
WORTH A SHILLING
FOR NOTHING!

DO IT NOW.

Fred Warren,
NORTHAMPTON

NATIONAL UNION OF
POLICE AND PRISON OFFICERS

Furniture, Bedding, etc.

CYCLES

Mortons

Fig. 1.2 The typical front cover of *The Police Review* published during the war

Britain became increasingly reliant on food imports so that British farming was considered unimportant on the outbreak of war and mainly involved pastureland grazing cattle. When the enemy blockaded food imports and the poor winter weather globally produced a lower wheat harvest internationally combined with a poor potato harvest in the UK, Britain faced the threat of national starvation.

Chapter 3 shows the deterioration of manpower levels in British agriculture before and since the outbreak of war, with the inability of farmers to effectively continue even with the same levels of food production without significant help. Previous government reports anticipating Britain's fate if imports were curtailed, set the scene for government control of agriculture in the national interest. With a change of Prime Minister in December 1916, government took control of farming to urgently increase crop production to feed the nation, overturning decades of increases in pasture to graze cattle at the expense of arable. The call to farmers from the President of the Board of Agriculture was "back to the 1870s" and gives his rationale for increasing arable cultivation at the expense of grazing pastureland. However, arable was known to have higher costs, needed more investment and had greater risks for a variable income than most pastoral farming [63]; therefore, farmers would need inducements to convert their pasture. This chapter shows the national systems established to provide farmers with manpower and supplies to appease them and encourage urgent and increasing levels of arable food production for human consumption until the end of the war, in attempts to make Britain self-sufficient in crops. This chapter demonstrates how government strenuously attempted to increase labour levels in agriculture from a large variety of sources, including policemen, so that no stone was left unturned, while increasingly protecting agricultural labour from recruitment into the army. This national emergency raised the population's concern for and interest in British farming.

Chapter 4 shows the mechanical and horsepower help given to farmers. Increased mechanisation attempted to minimise the need for unsustainable increases in manpower to meet the Board of Agriculture targets. We see the urgent beginnings of petrol-driven tractors combined with the more traditional steam ploughs and how the number of horses for ploughing needed to be upheld when the army requisitioning many. At a time when the majority of ploughing was manual pulled by horses, the latent skills of policemen combining their physique and strength to hold the ploughshares sufficiently in the ground to turn the soil to a sufficient depth to plant crops led farmers to value their help.

Chapter 5 attempts to assess why government prioritised self-sufficiency in crop production, maximising carbohydrate intake over increased meat and milk production due to the dietary preferences by most of the population for low-cost food with the highest calorific value. With high food price inflation, a diet of mainly carbohydrates provided the most energy and had been staple for around 80% of the population since at least the turn of the century, a trend which continued into the war years despite shortages of some foods, such as potatoes. This chapter shows the real threats of civil unrest over food availability and rising prices with attempts to justify government intervention to plough up unused and under-used land and grassland to grow crops, despite being more labour intensive than grazing cattle.

Chapter 6 gives the locations and approximate numbers of policemen released full-time into agriculture in England in 1917 and 1918 or in some areas they helped agriculture on a part-time basis. Details of each police force are given, where this is known, and the approximate proportions of each released into agriculture as well as the local geography which determined the types of farming. Chapter 7 identifies the differences between England and Scotland in agriculture management during the war and how Scottish police forces were approached for help. In both these chapters we see the care in selecting policemen with appropriate skills, the promise of continuation of their conditions of service and levels of pay, and the lengths of time over which each police force released their men with the conditions attached, such as working close to home to be recalled to undertake police duties if an emergency arose. Other sources of help to farmers during the second half of the First World War have become noteworthy, such as the Women's Land Army (see for example Clarke [64]; Scott [65]; Scott, [66]; and Charman [67]). However, most of this help was given after June 1917, by which time many policemen had returned to police duties. Therefore, the release of policemen to help farmers pre-dated large-scale release of other adult groups, excepting the army, and to date this help has not been recognised. This book addresses the gap in British social history which reveals the police as first responders.

Chapter 8 provides two case studies—Birmingham and Glasgow—where large numbers of policemen were released into agriculture, both acted as leaders in helping agriculture, with the particular local circumstances surrounding their release. Birmingham was chosen because it was the home of Neville Chamberlain, the ill-fated first Director of National Service, developed as a first attempt at national manpower planning, although on a voluntary basis. The city released around 100 policemen permanently from Spring 1917 until the end of the war; this is the only location found not to recall policemen during times of less critical need for agricultural manpower. Glasgow is chosen as the second case study, as it was heralded by other Local Authorities in Scotland as the pioneer and was targeted by Neville Chamberlain in his campaign to raise volunteers for National Service. It was also a location where near food riots occurred in Spring 1917; the reasons for this are shown along with the Corporation's response.

The final chapter summarises the release of policemen into agriculture and attempts to account for why Chief Constables agreed to such releases when regular police numbers had already changed dramatically, and regular manpower was so depleted due to recruitment into the war. The outcome of the wartime food production campaign is considered with its ability to feed and pacify the nation alongside the success of other combatant nations to feed their populations, showing how Britain's response largely overcame the food rioting seen in other warring nations. Also briefly considered is the continuing role of the British police as first responders in times of national crisis and disaster.

References

1 Rawlings, P. 2012. *Policing: A short history.* Taylor and Francis https://www.perlego.com/book/1514809/policing-a-short-history-pdf (accessed 11/1/2024) Chapter 5.

2 Weinberger, B. 1991. *Keeping the Peace? Policing Strikes in Britain, 1906–1926.* Berg.

3 Morgan, J. 1987. *Conflict and Order: The Police and Labour Disputes in England and Wales, 1900–1939.* New York: Clarendon Press of Oxford University.

4 Neocleous, M. 2021. *A Critical Theory of Police Power: The Fabrication of Social Order.* London: Verso p. 24.

5 Lawrence, P. 2016. Historical Criminology and the Explanatory Power of the Past. *Criminology and Criminal Justice* 19(4), 493–511. https://doi.org/10.1177/1748895818794237 (accessed 11/1/2024).

6 Loader, I. 2020. Revisiting the police mission. Insight Paper 2. *The Police Foundation* https://www.policingreview.org.uk/wp-content/uploads/insight_paper_2.pdf (accessed 11/1/2024).

7 Bierschenk, T. 2016. Police and State. In Bradford, B et. al. *The SAGE Handbook of Global Policing* 155–178 pp. 166–7.
8 Bauman, Z. and Bordoni, C. 2014. *State of Crisis.* Cambridge: Polity, pp. 1–12.
9 Lawrence, P. 2016b. The historiography of Crime and Criminal Justice. In *The Oxford Handbook of the History of Crime and Criminal Justice,* eds. Knepper, P. and Johansen, A. OUP pp. 17–37; p. 25.
10 Charman, S. 2017. *Police Socialisation, Identity and Culture: Becoming blue.* Palgrave Macmillan, pp. 337–8.
11 Fielding, N. 1986. Evaluating the Role of Training in Police Socialisation: A British Example. *Journal of Criminal Psychology* 14, July, pp. 319–30.
12 College of Policing 2015. *College of Policing Analysis: Estimating demand on the police service.* p. 9, https://www.college.police.uk/search?query=Estimating+demand+on+the+police+service (accessed 11/1/2024).
13 Gregory, A. 2008. *The Last Great War: British Society and the First World War.* Cambridge: Cambridge University Press.
14 Offer, A. 1991. *The First World War. An agrarian interpretation.* Oxford: The Clarendon Press, p. 1.
15 Lloyd George, D. 1934. *Memoirs* Volume 3. London: Ivor Nicholson and Watson p. 1269 https://archive.org/details/WarMemoirsOfDavidLloydGeorgeVolIii/page/n263/mode/2up?view=theater (accessed 11/1/2024).
16 Ziemann, B. 2014. Agrarian Society. In *The Cambridge History of the First World War,* ed. Jay Winter. Volume II The State. Cambridge University Press, Chapter 15, pp. 382–407.
17 Although this half-column advertisement is taken from *The Yorkshire Evening Post* on Friday 16 March 1917, p. 4, it was also placed prominently in most local and national newspapers for many days in the middle of March 1917(17).
18 *The Police Review and Parade Gossip.* Police for Farm Work. Question discussed at Glasgow. 9 March 1917, p. 75.
19 Philip, N. 1993. *Victorian Village Life.* Albion, pp. 113 and 137.
20 Royal Commission on the Police *Interim Report,* 1960 Cmnd. 1222 TNA HO 272.
21 *The Police Review and Parade Gossip,* Answers. May 19, 1905 p. 230.
22 Shpayer-Makov, H. 2002. *The Making of a Policeman: A Social History of a Labour Force in Metropolitan London.* Aldershot: Ashgate, pp. 25–94.
23 Emsley, C. 1996. *The English Police: A Political and Social History* 2nd Edition, London: Longman, p. 195 Table 9.2.
24 *Glasgow City Police Records* SR22/55/17–24. Available from The Mitchell Library Archives, Glasgow.
25 Fraser, M. 2018. "Police as Ploughmen": Temporary Release to Help Farmers in the Food Crisis of First World War Britain. *Cultural and Social*

History. Published online 5 August 2021. https://doi.org/10.1080/14780038.2021.1958443 (accessed 11/1/2024) 18(5), 651–667.

26 Klein, J. 2002. Blue-collar Job, Blue-collar Career: Policemen's Perplexing Struggle for a Voice in Birmingham, Liverpool, and Manchester, 1900–1919. *Crime, History and Societies* 2002, 6(1), 5–29.

27 Philip. 1993. op. cit. pp. 113 and 137.

28 Emsley, 1996, op. cit. p. 94.

29 Emsley, C., and M. Clapson. 1994. Recruiting the English policeman C. 1840–1940. *Policing and Society: An international journal* 3: 269–289. https://doi.org/10.1080/10439463.1994.9964675 (accessed 11/1/2024).

30 Emsley, 1996, op. cit. pp. 197–8.

31 Davidson, N., et al. 2017. Police and Community in Twentieth-century Scotland: The Uses of Social History. *The British Journal of Criminology* 57 (1): 18–39. https://doi.org/10.1093/bjc/azv097 (accessed 11/1/2024).

32 Jackson, L. *et. al.* 2020. *Police and Community in Twentieth-century Scotland*. Appendix: The origins of police recruits in Scotland pp. 218–224 Edinburgh University Press. https://doi.org/10.3366/edinburgh/9781474446631.001.0001 (accessed 11/1/2024).

33 Local Government Act of 1888 (51 & 52 Vict.) c.41.

34 Report of the Committee on the Police Service of England, Wales and Scotland, Part 1. Cmd. 253, 1919 pp. 4–5. The Desborough Report.

35 Emsley 1996 op. cit. pp. 197–8.

36 Cowley, R. 2011. *A History of the British Police: From its Earliest Beginnings to the Present Day*. The History Press p. 135.

37 Emsley, 1996, op. cit. pp. 121–2.

38 *The Police Review and Parade Gossip* March 23, 1917, p. 95.

39 *The Police Review and Parade Gossip* June 15, 1917, p. 188.

40 Surrey County Council Archive Centre CC/98/22/1 in Bartlett, R. ed. July 2022. *In Blue Serge or Khaki; always police officers*. Surrey Constabulary History https://www.surrey-constabulary.com/ (accessed 25/1/2024).

41 The Police (Emergency Provisions) Act 1915 (5 & 6 Geo. 5: Ch. 41).

42 The Police Review and Parade Gossip 26 January 1917 p. 28.

43 Emsley, 1996, op. cit. pp. 123–5.

44 Emsley, 1996, op. cit. pp. 123–9.

45 Emsley, C. 2009. *The Great British Bobby: A History of British Policing from 18th century to the present*. London: Quercus pp. 188–91.

46 Fraser, M. 2019. *Policing the Home Front: The Control of the British Population at War*. Abingdon: Routledge, Chapter 3. Controversies over the War Separation Allowances, pp. 29–62.

47 *The Police Review and Parade Gossip* 15 September 1916 p. 393.

48 Macmillan, M. 2021. *For the Love of History: A Conversation with Margaret MacMillan*. Ideas Roadshow Open Agenda Publishing ISBN 978-1-77170-146-4 https://books.google.co.uk/books?id=ZFcsEAAAQBAJ&printsec=frontcover#v=onepage&q&f=false (accessed 11/1/2024).
49 Ben-Porath S.R. 2018. Wartime Patriotism. In *Handbook of Patriotism*, ed. Sardoc M. Springer, Cham. pp. 693–706. https://doi.org/10.1007/978-3-319-30534-9_3-1 (accessed 11/1/2024).
50 House of Commons debates HC Deb 27 April 1915 vol 71 c553.
51 *The Police Review and Parade Gossip* Police Law and Practice: Questions and Answers. 13650. National Registration Act—*Police Powers* (italics supplied). 14 July 1916, p. 313.
52 Charman, T. 2014. *The First World War on the Home Front*. London: Imperial War Museum p. 55.
53 Local Government Board 1916. Circular R 97 from the Local Government Board to Local Tribunals and Appeal Tribunals 26th August 1916. TNA MH 47/142/4/1.
54 Monger, D. October 2018. Tangible Patriotism during the First World War: Individuals and the Nation in British Propaganda. *War and Society* 37(4), 244–61. https://doi.org/10.1080/07292473.2018.1496786 (accessed 11/1/2024).
55 *The Police Review* Parade Gossip. 8 December 1916, p. 494.
56 *The Police Review* Discontent in Pembrokeshire. 26 January 1917 p. 28.
57 Emsley, 2009, op. cit. pp. 172–5.
58 Emsley, 1996, op. cit. pp. 101–103.
59 Fresno, A. *The Times*, 15th January 2010 p.
60 Clapson, M. and Emsley, C. 2011. Street, beat and respectability: The culture and self-image of the late Victorian and Edwardian urban policeman. In *Police and Policing in the Twentieth Century*, ed. Williams, C. A. Farnham: Ashgate Publishing Chapter 14, pp. 293–317.
61 Klein (2002) op. cit. pp. 8–17.
62 Silverman, D. 2021. *Doing Qualitative Research* 6th edition London: Sage p. 300.
63 Whetham, E. H. 1974. The Agriculture Act, 1920 and its Repeal—the "Great Betrayal" *The Agricultural History Review* 22(1), 38.
64 Clarke, G. 2008. *The Women's Land Army: A Portrait*. Bristol: Sansom.
65 Scott, I. 2007. The Land. In *Women Workers of World War I*, ed. Stone, G. ed. Chapter II pp. 46–66. Mansion Field.
66 Scott, C. 2017. *Holding the Home Front: The Women's Land Army in the First World War*. Barnsley: Pen and Sword.
67 Charman, T. 2014. op. cit.

CHAPTER 2

The Developing Food Crisis in the First World War

Introduction

This chapter gives the social and political context resulting in the temporary release of policemen into agriculture from the beginning of 1917. It starts by showing the laissez faire development of British agriculture with the continual movement of populations away from rural areas, Barnett says that whereas in 1871 over 1 million workers were agricultural labourers in England and Wales, by 1914 only 668,000 male and 36,000 female workers remained [3]. This trend developed from grain and other easily transportable food products being grown more cheaply overseas and led to changes in British farming from a decrease in arable land to an increase in pastureland to graze cattle, requiring less manpower. In addition to their dwindling permanent workforce, farmers became accustomed to using temporary labour in times of heavy workload, such as spring and autumn and had built up regular sources of supply. Britain increasingly relied on imports of foodstuffs, particularly grain, to sustain the population, as the importance of farming as an occupation lessened and landowners diversified into other sources of income, while industry in the towns and cities rose in importance, providing work for many. From at least the beginning of the 1900s Britain understood the threats to feeding the population if there were reduced imports of food and food products and prepared for such an event [4]. How this played out during the First World War would include progressively tight government control over

M. Fraser, *The British Police and Home Food Production in the Great War*, https://doi.org/10.1007/978-3-031-58743-6_2

national labour sources including substituting labour into agriculture and the protection of farm labour from conscription.

This chapter demonstrates the sharp contrast at the beginning of 1917 compared with peacetime and the first two years of the First World War. It shows the population's anxiety about feeding the Home Front, with the threat of social disorder, following the severe reduction in food imports due to blockading and the poor national and international harvests in 1917, the year of war weariness.

The ability of British agriculture to feed the population before the Great War

From around 1875 changes in British agriculture started in the sixteenth century [5] continued. The mainly prosperous years of the mid-nineteenth century gave way to agricultural changes which saw long periods of falling prices for produce grown by tenant farmers and owner occupiers of small farms, lower rents for landowners, untenanted farms and increasingly disused and under-used land.[1] Associated with agricultural decline was the demise of many manufacturing industries based in rural areas. However, despite this decline and rapid population growth, Britain could still feed 80% of its people from home grown produce [6]. In the towns and cities manufacturing rose in prominence as Britain led the world in industrialisation, while the population depended on ever-increasing imports of food and with reduced costs of transportation, this stimulated other nations to improve their agricultural production to supply Britain which they could produce more cheaply [7]. British farming ceased to be a major industry with over half a million people leaving the farms. Whereas in 1851 agriculture employed around one-quarter of adult males in work, by 1901 the proportion had dropped to between 12 and 16% [8, 9]. Further migration of agriculturalists to non-European countries occurred between 1912 and 1914 [10]. For the UK as a whole, agricultural output declined by around 0.2% every year between around 1875 and 1900 but rose by around 0.3% between the late 1890s and the outbreak of the First World War [11]. However, output in England and Wales was regionally specific and increased in the areas

[1] The pre-war conditions for small family-run tenanted farms are depicted in the opening scenes of the Stephen Spielberg 2011 film War Horse.

devoted to grazing, mainly in the west, north and midlands, while falling in the arable areas of the south and east [12]. Associated with this decline was wealthy landowners' loss of social and political influence [13]. From the 1880s many landowners sold or attempted to sell all or parts of their estates and many experienced financial difficulties [14], others diversified into different sources of income, while the public became apathetic about British farming [15]. However, by 1914 agriculture was still the largest single employer and the largest user of land [16] despite farm workers' wages being lower than other occupations throughout the period 1850–1914 [17]. Only from 1916 with the impact of food shortages and soaring food prices was concern raised for British agriculture, when the War Emergency Workers' National Committee claimed to speak for the entire working and lower middle classes by campaigning and brought food availability, prices and distribution to the fore. By mid-1917 their influence was felt by government [18].

Decades of changes in farming can largely be assigned to laissez faire government policies of free trade, free currency and free enterprise [19]. Britain was no longer able to feed the population from home-grown produce and became reliant on imported food and food products such as wheat [20], which, in a largely pre-mechanised era, required large amounts of skilled labour to plough the land with hand ploughs pulled by horses. By 1900, with improvements in marine technology, Britain imported nearly 60 million tons of grain annually mainly from North America, Canada and Argentina. Between around 1875 and 1905 the quantity of wheat grown in Britain declined by 50%, while consumption rose by 300%. The Royal Commission published in 1905 [21] estimated that Britain had supplies of wheat to last only six weeks at current levels of consumption. Britain's vulnerability to loss of food imports and raw materials by enemy blockade in times of war and the consequent rise in prices had been shown since 1903 in the 1905 Royal Commission Enquiry [22].

Before the war writers on British agriculture revealed the differing acreage and ratios given over to pasture and arable, year on year. Rowland Prothero, later raised to the peerage as Lord Ernle, and appointed President of the Board of Agriculture from December 1916, in his influential text *English Farming Past and Present* [23] said from 1909–13, 36 million acres were devoted to rearing livestock, while only 3 million acres grew crops such as grain and potatoes. But these figures were fiercely contested by Kerridge [24] as not being based on historical facts, so they need to be treated with caution. By 1914 Middleton's rather self-congratulatory text showed an increase in acreage and a different distribution at 27.3 million acres of permanent grassland and 19.5 million acres of arable [25], with which the agricultural statistician Henry Rew, agrees [26]. By the outbreak of war, agricultural land use was 60% pasture grazing cattle, as it was more profitable for farmers than using land to grow crops [27, 28]. By 1915 with government advice to increase arable, 34 million acres were said to be given over to grass with only 13 million acres to arable [29]. These figures demonstrate how British agricultural land had given way to pastoral farming, requiring a vastly reduced labour force [30]. A broad division of British land use was between arable farming in the south and east and pastoral in the north and west, although most farms were mixed use.

Although the reliance on food imports was known, it was not until 1916 that the Royal Society showed the extent by estimating average calorific intake: the UK bought the largest amount of wheat in the world, importing 81% of its calorific intake from which Britain made bread. These imports accounted for three-quarters of the surplus stock of wheat imported into Europe and Brazil [31]. Imports supplied the British population for 42 weeks of the year, whereas home-grown wheat only supplied bread for 10 weeks [32–34]. Imports of other cereals accounted for 32% of consumption, whereas imports of meat were around 40%. However, Britain produced all the milk supply and 94% of potatoes, making their home production crucial [35].

Feeding the Population 1914–1916, during the Asquith Government

By the outbreak of war, Britain imported around 60% of all calories needed to feed the population; only 16 million people were fed from home grown produce, 37.2% of the population [36–38]. This prompted the House of Lords to discuss wheat production and stocks in Britain as an urgent priority, but no further action was deemed necessary. Wheat growing, inducements to farmers to increase production, and wheat availability imported into Britain was also debated in the House of Commons on 9th September 1914 with similar reassurances that no further action was needed, as imports were secure [39, 40].

While imports remained plentiful, with bumper cereal harvests in Canada, USA and Argentina in 1915–16 [41] there was no recognised need for change; however, Britain's vulnerability if the enemy operated a blockade was acknowledged and led to the establishment of an Agricultural Consultative Committee which issued press releases advising farmers that they should increase home food production, particularly wheat and other cereals [42]. In the meantime, since the outbreak of war, efforts were made by government to build the nation's stock of wheat and to secure imports of meat and sugar [43], the latter entirely imported and an important foodstuff and source of energy [44]. Supplies of sugar were secured from August 1914 by the establishment of a Cabinet Committee on Food Supplies, creating a Royal Commission on Sugar Supplies chaired by Captain Charles Bathurst which took over the sugar trade [45–47]. These moves heralded government intervention in food, not only to provide for Britain but also for its troops in France and Italy. However, food prices started to rise, alarming and disadvantaging the working classes including the trade unions, causing social unrest by strike action into February 1915, although price rises were not sufficient to change dietary habits at this time [48].

Further developments to feed the population in wartime came with the appointment of Lord Selborne, Minister of Agriculture from May 1915 in the first Asquith coalition government. He established a committee chaired by Lord Milner to consider food production in England and Wales, a similar committee was established in Scotland, chaired by Eugene Wason M.P. [49]. The interim Milner report of June 1915 [50] concluded that a "plough policy" was needed to restore poorer grass and land uncultivated since the 1870s to bring it back into cultivation. The report recommended farmers increase their arable land by at least one million acres in 1916 and more in later years, either by increasing the area under wheat by one fifth more than in October 1913, or by converting at least one fifth of the current area under grass into arable, showing the economic benefits to farmers [51], while government should raise the prospect of paying farmers a minimum price for wheat for four years as an incentive [52–54]. Further recommendations were that landlords should not raise farm rents so that farming remained an attractive occupation, that farmers should offer higher wages to farm labourers and that each county should establish a committee to offer advice on the most suitable crops to grow. Each county committee should appoint district committees to work closely with farmers to inform them how much arable land had been lost since the 1870s and the extent of tillage required for the next harvest [55, 56]. The Wason Committee made similar recommendations in Scotland in their first report of May 1915 but rejected a guaranteed price for cereals [57, 58], instead arguing for more Liberal feeding of cattle, making Scotland's main contribution in meat and milk, but this was later overturned [59]. Consistent with Liberal laissez faire policies, the report recommended advice to farmers rather than direct intervention. But both reports were side-lined in Cabinet on 5th August 1915 [60–62] as the threat of diminished imports was insufficient and world cereal harvests were good. However, the Board of Agriculture and Fisheries in England and Wales and the Board of Agriculture for Scotland established a non-mandatory system of War Agricultural Committees (WACs) at local authority level from autumn 1915, they were to consider manpower data from the National Register[2] and, with a Treasury grant, organise the local supply of labour, consider how food production could be maintained and increased and report on shortages of supplies including fertilisers, animal feed, machinery and other items such as seed [63–65]. WACs were advised to establish district committees to provide direct access to farmers and report on the adequacy of labour to the WAC, this would provide the Boards with detailed national knowledge. Without executive powers, WACs established channels of communication between the counties and the Boards. But not every local authority set up a committee; those established were unwieldy due to large memberships, did not represent farm labourers and were inefficient [66, 67]. The Scottish report also agreed with the establishment of district committees to offer advice to farmers and to provide government with information on farmers' needs for labour and supplies, these were initiated by the Scottish Board of Agriculture by autumn 1915 [68]. The Agricultural Consultative Committee with a UK membership also sent notices to the press addressed to farmers in August and September advising them to increase production, particularly of wheat and other crops [69].

Despite government side-lining the reports, members of the Milner committee, including Selborne, continued to advocate the plough policy [70], due to threats to food imports not only by enemy action but by

[2] By mid-1915, with 75,000 men killed and the initial enthusiasm for volunteering for the army fading, government felt the urgent need for statistical information to identify all men and women between the ages of 15 and 65 years in civilian life, as different sets of figures were being used by the Board of Trade and the War Office to determine the future of trade and the size of the army. The National Registration Act passed on 15th July 1915 aimed at holding a register of all civilians for the duration of the war; it collected data on marital status, address, date of birth, number of dependents, current employment and other skills than the current job.

loss of shipping if the army requisitioned more commercial ships to carry troops and supplies, as by December 1915 they had withdrawn nearly a third of commercial shipping [71, 72]. Selborne's proposals on 23 March 1916 to the War Committee to increase home food production gave 2 options:

1. Non-compulsion would require offering farmers a guaranteed price for wheat and assistance with farm labour; or
2. Compulsion would require the Board of Agriculture to have legal powers to increase funding to the WACs; to require surveys of the land to establish where ploughing would be to the best advantage and to provide farmers with labour and machinery with the need to consider agricultural wages [73, 74].

But these proposals were fiercely opposed in the War Committee, notably by Arthur Balfour, First Lord of the Admiralty who interpreted them as implying the admiralty had lost control of the German submarine threat [75], and by Treasury Minister Reginald McKenna and Walter Runciman at the Board of Trade. The proposals were not discussed again until May 1916 when it was also agreed to take no further action [76, 77], which can be understood as food production in the UK rose in 1914 and 1915 above pre-war levels but fell in 1916 [78]. However, Selborne was sufficiently alarmed at labour loss from farming that he wrote to WACs in May 1916 imploring them to urgently ask their district committees to plan for the number of replacements needed for the 1916 harvest, suggesting they recruit from men over military age and from schools [79]. Relaxing school attendance for boys and girls in local authority elementary education started from the outbreak of war, particularly occurring in rural areas. The same did not occur for children in public and secondary schools [80]. The different and diverse sources of manpower help for farmers is shown in Chapter 3.

While imports were maintained and farmers appeared to be carrying on as usual with a depleted workforce, manpower losses in agriculture did not cause undue concern in government and the level of substitution to help farmers was said to be adequate to feed the population, while food shortages from enemy action on imports were negligible up to autumn 1915 [81, 82]. However, with an increasing number of ships being sunk by enemy action from 18th October 1916, the loss of imports inevitably increased. The loss of ships escalated from October 1916, so that by January 1917 the total loss of all types of ships was 175. This led to agreement by the War Cabinet on 13th November to appoint a Food Controller, Lord Devonport was appointed on 8th December [83, 84]. In Scotland, the threat of submarine warfare led to recall of the Wason Committee in November 1916 to consider how arable production could be increased.

Despite advice to increase home food production, by the end of 1916 available agricultural land was slightly reduced compared with 1915 with a decrease in the area devoted to crops of 148,000 acres, although increased acreage of wheat and oats were grown. However, other grain crops and potatoes decreased in acreage over 1915 [85]. If food imports continued to fall, increases in agricultural production and food preservation measures would be needed urgently to feed the population.

Feeding the Population During the Lloyd George Government: December 1916 to the End of the War

With the change of government in December 1916, Lloyd George was appointed head of the coalition. The period heralded increasing fears of widespread population starvation across Britain, which had emerged as early as the 1890s in popular fiction and as questions in the House of Commons [86]. Advice was no longer seen as adequate and highly interventionist policies involving government control of agriculture were rapidly implemented, reaching from central government down to the individual in rural areas. However, government support was not automatically obtained, as agriculture was the last industry to be considered after the army, navy and munitions, which were backed by strong public support. Agriculture had "not only to arouse farming opinion, but to make the urgency of its needs felt by the Press, by politicians and by the people" [87]. The established systems aimed at rapid development of British agriculture to feed the population from home grown food while keeping farmers on side; policies focussed on crop production with decreased reliance on food imports. Furthermore, by early 1917 Britain acknowledged that total war meant every citizen must be involved, transforming the country "into a war machine of the whole national organisation" [88].

The national food emergency became evident from early January 1917 when grain stocks in Britain were declining, allowing Germany to predict a knock-out blow could be struck by increasing attacks on merchant shipping [89]. Germany planned to end the war speedily due to its own threats of population starvation from a poor harvest in 1916; Germans were reduced to living on 1,000 calories per day, insufficient to maintain health [90]. Fears of starvation in Britain increased from 1st February, following the declaration by the German government of unrestricted submarine warfare and their rapid increase in the number of submarines to further blockade imports, this resulted in an increase in the total number of ships sunk to 136 in the first two weeks of February alone, estimated to increase to 238 vessels for the entire month of February [91]. The prospects and, within a month, the reality of decreased food imports came just 3 months after the end of the battle of the Somme, with the largest loss of life in the war to date. This direct blow of food insecurity on the Home Front contributed to demoralisation and war weariness.[3]

Lloyd George told parliament on 23rd February that the war had reduced the amount of shipping by 20,000,000 tons, used to service the war effort in France, America, Egypt and India [92]. The shipping crisis reached its height in March and April 1917 as losses continued to rise, these were widely publicised in the press:

> SHIPPING LOSSES
> THE TOLL OF SUBMARINES AND MINE.
> ANOTHER HEAVY LIST
> The Admiralty return of arrivals, sailings, war losses, and unsuccessful attacks for the week ended 3p.m., April 8, 1917, was issued last night and is given below, together with the figures for the previous two weeks.

British merchant vessels sunk by mine or submarine		April 8	April 1	Mar. 25
(a)	1,600 tons (gross) or over	17	18	18
(b)	Under 1,600 tons (gross)	2	13	7 [108]

The quotation, published in *The Times* for a national readership, gives the alarming total of consistent weekly loss of ships by enemy submarines. These frequent reports heightened national anxiety and war weariness. Between 1st and 18th April, Cabinet reports told of 175 ships sunk [93],

[3] https://encyclopedia.1914-1918-online.net/article/civilian_morale/ See the paragraph "Civilian Morale: National Cases" para 2.

outstripping the construction and purchase of new ships and the repair of damaged ships. Newspaper reports throughout Britain continued to show the seriousness of the situation for food supplies, for example *The Times* 14th April 1917 p. 7;[4] similar articles were seen in the *Leicester Daily Post* 20th April 1917 p. 4 and in other national and local newspapers. However, questions in the House of Commons indicated suspicion that the Admiralty was not making the entire situation known for fear of informing the enemy [94], so that the War Cabinet agreed on 9th May 1917 the number of ships sunk by enemy submarines would be published after a period of two weeks had elapsed [95]. In May 1917 reports to the War Cabinet showed Germany was concentrating all its shipbuilding resources into the construction of submarines [96].

However, it was not only the availability of imports that threatened British food supplies. The blockade was combined with the poor international harvest of 1916 due to adverse weather conditions. Failure of winter wheat in America, Canada and Argentina, Britain's main suppliers, produced the smallest crop for 20 years so that exports were threatened, in addition to a smaller crop of wheat in Britain [97–99]. Indeed, lack of international control of wheat exports led to riots in New York in mid-February 1917, when shops ran out of bread [100].

In Britain the severe autumn and winter weather of 1916/17 also reduced areas of British winter-sown wheat and caused a serious shortage of home-grown potatoes evident in the shops and markets[5] from January 1917 which lasted for many months [101, 102]. The poor weather reduced the British potato crop by around 16%, with the Scottish crop decreased by around a half from pre-war levels. In addition to bread, potatoes were an important and cheap food source for the masses, on which they relied. The third major food shortage was sugar, an important source of energy in the British diet. Supplies were entirely imported and declined by around one-sixth from pre-war levels [103].

In Chapter 5 I consider how these shortages particularly affected the working classes, around 80% of the population [104]. Less than 20% of people over 10 years of age were classified as middle or upper class

[4] The article was titled "The dearth of wheat. Bread economy essential. The King's example in rationing. Smaller herds." Followed later in the article by the heading "Big decline in world wheat supply. Urgency of bread saving."

[5] The importance of bread and potatoes in the national diet is shown in Chapter 5 and the effects of shortages on the population are seen in Chapter 8 in Glasgow.

according to the 1911 census[6]; with higher incomes they would be less affected by shortages and price rises, making the diets of the masses a serious source of concern [105].[7] The dietary habits of the population by income levels are developed further in this later chapter.

Early attempts to increase Britain's self-sufficiency in crops to feed the nation were hampered by the poor weather conditions, which delayed ploughing and planting crops for a 1917 harvest until mid-March [106]. The very unfavourable wet weather in February and March 1917 froze the ground making ploughing arduous, so that planting could not start until late March or mid-April [107]. From mid-March, with a break in the long harsh winter, the urgency to plough and sow crops saw resumed intense activity on the heavy, wet ground; this was when policemen were first seen released into agriculture to help farmers in the national crisis.

The next chapter details the increasing urgency of supplying labour into agriculture and the multiple sources of labour in attempts to help farmers provide the population with home grown food during the last two years of the war. It shows the urgency which required government to adopt compulsory measures with farmers, so that they worked together under the revised War Agricultural Executive Committees (WAECs) to produce a maximum harvest from arable land. The chapter shows how agricultural compulsion worked at local authority and government levels by surveying the land and issuing farmers with detailed instructions on which fields to plough and the crops to grow, under threats of sanctions if they did not comply. Recognising that they needed help not only with large increases in manpower but also with equipment, fertilisers, seeds and finance, a whole system was established under the Food Production Department to help farmers.

[6] These 20% included working in either general or local government, defence of the country, a professional occupation, a commercial occupation or those with private means.

[7] The census also included the populations of Scotland and the whole of Ireland, before partition, and is included in the estimates.

References

1 Journal of the Board of Agriculture, April 1915—March 1916 https://archive.org/details/dli.ernet.232728/mode/2up (accessed 25/1/2024).
2 Journal of the Board of Agriculture, April 1917—March 1918 https://archive.org/details/dli.ernet.29517/mode/2up (accessed 25/1/2024).
3 Barnett, M. L. 1985. *British Food Policy During the First World War* Boston: George Allen and Unwin p. 4.
4 Barnett, 1985, op. cit. p. xv.
5 Kerridge, E. 2013. *The Agricultural Revolution.* London: Routledge. https://doi.org/10.4324/9781315019888 (accessed 15/1/2024).
6 Mingay, G. E. 1963. The "Agricultural Revolution" in English History: A reconsideration. *Agricultural History* 37(3), 123–33.
7 Offer, A. 1989. *The First World War: An Agrarian Interpretation.* Clarendon Press, pp. 81–5.
8 Dewey, P. E. 2014. *British Agriculture in the First World War.* London: Routledge. 2nd edition, p. 3.
9 Collins, E. J. T. 2000. Introduction and Chapter 2 Rural and Agricultural Change. In *The Agrarian History of England and Wales,* ed. E. J. T. Collins VII. Part I. 1850–1914. Cambridge: Cambridge University Press, pp. 1–29 and 204–7.
10 Dewey, P. E. 1984. Military recruiting and the British labour force during the First World War. *The Historical Journal* 27(1), 199–223, p. 223.
11 Collins, 2000, op. cit. p. 204.
12 Collins, 2000, op. cit. pp. 217–23; Tables 2C.5 and 2C.6.
13 Mingay, 2000, op. cit.
14 Beckett, J. V. 2000. Agricultural landownership and estate management. Chapter 10. In *The Agrarian History of England and Wales,* ed. E. J. T. Collins. VII (Part 1) 1850–1914, pp. 717–8.
15 Ernle, Lord. n.d. *The Land and its People; Chapters in Rural Life and History.* London: Hutchinson, p. 121.
16 Collins, 2000 op. cit. p. 5.
17 Dewey, P. 2000. Chapter 12 Farm labour. In *The Agrarian History of England and Wales,* ed. E. J. T. Collins. VII (Part 1) 1850–1914. Cambridge: Cambridge University Press, pp. 820–862.
18 Barnett, 1985, op. cit. pp. 131–2.
19 Marwick, A. 2006. *The Deluge: British Society and the First World War.* Basingstoke: Palgrave Macmillan. Reissued Second Edition, pp. 191–2.
20 Barnett, 1985, op. cit. pp. xiv and 3.
21 *Royal Commission on the Supply of Food and Raw Material in Time of War*, pp. 1905, xxxix Cd 2643 Volume 1; The Report.

22 Royal Commission on the supply of food and raw material in time of war, p. 7 paras 37–8; see also the debates since 1909 on Britain's food supply in times of war in Barnett, 1985, pp. 9–19.
23 Prothero, R. E. 1961. *English Farming Past and Present,* 6th edition. London: Heinemann, p. 394.
24 Kerridge. 1969. op. cit.
25 Middleton, T. H. 1923. *Food Production in War.* Oxford: The Clarendon Press, p. 14.
26 Rew, R. Henry. January 1918. The prospects of the World's Food Supplies after the War. Paper read to the Royal Statistical Society December 18, 1917, *Journal of the Royal Statistical Society* 81(1), 41–74.
27 Dewey, 2014, op. cit. p. 11.
28 Ernle, n.d., op. cit. p. 120.
29 Middleton, 1923, op. cit. p. 106.
30 Perry, P. J. ed. 2006. *British Agriculture: 1875–1914 (1st edition).* Routledge. https://doi-org.ezproxy.lib.gla.ac.uk/10.4324/9781315019109, pp. xi–xxiii.
31 Rew, 1918 op. cit.
32 Ernle, n.d., op. cit. p. 151.
33 McDermott, J. 2011. *British Military Tribunals, 1916–1918: A Very Much Abused Body of Men.* Manchester Scholarship Online https://academic.oup.com/manchester-scholarship-online/book/13569 (accessed 25/1/2024), p. 95.
34 Rew, 1918 op.cit.
35 Dewey, 2014, op. cit. p. 16 and ref 19 p. 19.
36 Dewey, 2014, op. cit. p. 4 and Table 2.7 p. 16.
37 Ernle, n.d., op. cit. p. 108.
38 Office for National Statistics https://www.ons.gov.uk/peoplepopulationandcommunity/populationandmigration/populationestimates/adhocs/004356ukpopulationestimates1851to2014 (accessed 27/3/2024).
39 HL Deb 27 August 1914 vol 17 cc520–31.
40 HC Deb 09 September 1914 vol 66 cc553–4 in Whetham, E. H. 1978. *The Agrarian History of England and Wales* Volume VIII 1914–39. Cambridge: Cambridge University Press p. 70.
41 Middleton, 1923. op. cit. pp. 135–6.
42 Dewey, 2014 op. cit. p. 23.
43 Beveridge, W. H. 1928. *British Food Control,* London: Oxford University Press. pp. 6 and 10–15.
44 Whetham, 1978, op. cit. p. 74.
45 Dewey, 2014, op. cit. p. 24.
46 Journal of the Board of Agriculture, 1915, pp. 517–518.
47 Rew, 1918, op. cit. p. 70.

48 Barnett, 1985, op. cit. pp. 28–44.
49 Conacher, H. M. 1926. Increased Food Production. Chapter IX. In *Rural Scotland During the War* , eds. Jones, *et. al.* London: Humphrey Milford, Oxford University Press. pp. 122–86.
50 Milner Report. *Interim Report of the Departmental Committee appointed by the President of the Board of Agriculture and Fisheries to consider the Production of Food in England and Wales.* 1915. Cd. 8048 Parliamentary Papers 1914–1916, V.
51 *The Economic Journal* 1916 26(101) pp. 105–113.
52 Dewey, 2014, op. cit. p. 25.
53 Middleton, 1923, op. cit. pp. 119–121.
54 Whetham, 1978, op. cit. pp. 96–7.
55 Whetham, 1978, op. cit. p. 76 and 91.
56 Middleton, 1923, op. cit. p. 122.
57 Whetham, 1978, op. cit. p. 77.
58 Middleton, 1923, op. cit. p. 123.
59 Conacher, 1926, op. cit. pp. 163–4.
60 Whetham, 1978, op. cit. p. 77.
61 TNA CAB 37/132/3.
62 Dewey, 2014, op. cit. p. 6.
63 Journal of the Board of Agriculture, 1915, pp. 498–9 and 702–4.
64 Dewey, 2014, op. cit. p. 26.
65 Middleton, 1923, op. cit. p. 137.
66 Dewey, 2014, op. cit. p. 27.
67 Barnett, 1985, op. cit. pp. 64–5.
68 Conacher, 1926, op. cit. p. 165.
69 Middleton, 1923, op. cit. pp. 104–5.
70 Middleton, 1923, op. cit. p. 137 and 167.
71 Dewey, 2014, op. cit. p. 28.
72 Barnett, 1985, op. cit. pp. 69–70.
73 Dewey, 2014, op. cit. p. 29.
74 TNA CAB 24/2/19 pp. 2 and 5.
75 Journal of the Board of Agriculture, September 1915, p. 495.
76 Dewey, 2014, op. cit. p. 31.
77 Barnett, 1985, op. cit. p. 64–5.
78 Dewey, 1978, op. cit. pp. 83–4.
79 Journal of the Board of Agriculture, May 1916, p. 192.
80 Horn, P. 1983. The employment of elementary schoolchildren in agriculture, 1914–1918. *History of Education* 12(3), 203–215.
81 Rew, 1918, op. cit.
82 Journal of the Board of Agriculture, 1915 p. 516.
83 Barnett, 1985, op. cit. p. 86–94.

84 War Cabinet 13th November 1916 TNA CAB 42/24/5.
85 Middleton, 1923, op. cit. p. 154 Table XV and p. 190.
86 Barnett, 1985, op. cit. p. 5.
87 Ernle, n. d. op. cit. p. 105.
88 The War Cabinet. 1918. Report for the year 1917. Cd. 9005 HMSO. https://archive.org/details/reportforyear19100grea/mode/2up?view=theater, (accessed 25/1/2024) p. 80.
89 Barnett, 1985, op. cit. p. 102.
90 Afflerbach, H. 2022. *On a Knife Edge: How Germany Lost the First World War.* Cambridge: Cambridge University Press.
91 Naval Policy in Relation to Mercantile shipping losses from Submarine Warfare, and the effect on the Strategical Situation. Memorandum by Lord Jellicoe, First Sea Lord, Appendix III. TNA CAB 24/6/40.
92 HC Deb 23 February 1917 vol 90 cc1592–3.
93 TNA CAB 23/2/35, Appendix; TNA CAB 23/2/43 Appendix I and II.
94 *HC Deb 27 March 1917 vol 92 cc205–6.*
95 TNA CAB 23/2/53 p. 2.
96 The War Cabinet 147, 25th May 1917 p. 2 TNA CAB 23/2/65.
97 Beveridge 1928, op. cit. p. 105.
98 Rew, 1918, op. cit. pp. 67–8.
99 Middleton, 1923, op. cit. pp. 160–161.
100 Barnett, 1985, op. cit. p. 172.
101 Middleton, 1923, op. cit. p. 154 and 161.
102 Whetham, 1978, op. cit. p. 90.
103 Middleton, 1920, p. 10, ref. 17 in Dewey, 2014, p. 35.
104 Barnett, 1985, op. cit. p. 3.
105 Census of England and Wales 1911. Summary Tables Cd. 5705 10 Ed 7 and Geo 5 Ch 27.
106 Ernle, n.d. op. cit. p. 101–3.
107 Middleton, 1923, op. cit. pp. 188–9.
108 *The Times* Apr 12, 1917, p. 7.

CHAPTER 3

Help for Farmers; No Stone Left Unturned

Help for farmers to increase home food production: No stone left unturned

Following from Chapter 2 setting out the developing crisis in Britain's national food supply, this chapter details the systems and national manpower crises within which the police worked during the second half of the war. It starts by showing how government increasingly encouraged farmers to expand crop production to feed the population by providing labour, although most was temporary and some of dubious quality. It highlights the progressive struggle by farmers to retain existing agricultural manpower against claims by the army for fit young men when, until late 1917, the army had priority for the national workforce over other essential industries on the Home Front. This chapter reveals the increasing use of substitutes in agriculture and their utility and reliability to farmers. Despite their own reduced numbers, the police were one of the first and most successful groups in early 1917 to be released from their everyday work to help farmers, a story that to date remains untold. The details of their release from each police force are shown in Chapters 6–8.

When the war did not end by Christmas 1914, the clamour of voluntary recruitment attributed to a combination of patriotism, a sense of adventure and escape from poverty and unemployment, which initially

M. Fraser, *The British Police and Home Food Production in the Great War*, https://doi.org/10.1007/978-3-031-58743-6_3

swamped the recruiting machinery, dwindled by early 1915[1] [3–5]. Initial recruitment from agriculture shows landowners' zealous encouragement to sign up by offering their sons and employees, some taking them to recruiting stations to enrol [6, 7]. Agricultural manpower in early 1914 was around 250,000 full-time male farmers and 700,000 full-time hired men working approximately 300,000 large farms, with additional part-time holdings mainly run by the farmer and his family [8]. Ernle estimated around 292,000 small farms, below 50 acres, 67% of farms in England and Wales [9], while agricultural statistics for 1914 showed 435,000 holdings of more than one acre in England and Wales with around one fifth containing less than 5 acres; under half were between 5 and 50 acres and only 3% were more than 300 acres [10]. Work by labourers on larger farms was specialised into skilled and unskilled; work with animals received higher pay, with horsemen commanding the highest wages due to their skill in controlling horses for ploughing, harvesting and carting, requiring a robust physique, which incidentally also made them the most attractive for army recruitment [11]. Farm work not involving animals was considered unskilled with lower wages. On smaller farms needing fewer staff, workers were versatile [12, 13].

From 1915 although farmers preferred to retain their skilled men of strong physique from army recruitment, signing up was uncontrolled leading to indiscriminate loss from this largely non-unionised industry [14]. In May 1915 the War Office instructed recruiting officers not to accept skilled farm workers,[2] to protect agriculture, but the instruction was often ignored [15–18[3]].

The struggle for manpower between the army and industries on the Home Front increased as war progressed, extending the need for more government control to protect essential industries. Initial attempts to raise the recruitment age to 40 and lower the minimum height for the infantry to 5 ft. 2 ins. failed to increase army recruits sufficiently [19] while labour losses from agriculture were significant. Thereafter, government reluctantly adopted conscription from March 1916. However, the police

[1] How this affected the police is shown in Chapter 1.

[2] The skilled farm workers included bailiffs, foremen, head carters, horsemen, stockmen, wagoners, shepherds and "necessary" milkers.

[3] Crowe shows how recruiting sergeants targeted men at the hiring fairs with patriotic messages about signing up for the "chance of their lives to join this glorious war" and farmers were asked not to obstruct recruiting.

were never an occupation protected[4] from conscription and experienced regular comb-outs.

The initial two years of escalating labour loss from the land was patchy [21, 22], for example upland areas of England and Wales, dominated by small family farms, experienced greater losses, while farmers in Devon felt unable to respond to demands for increased crop production in 1916 due to labour shortages and refused to cooperate with government until labour was assured [23]. Nationally, farmers became disheartened at their loss of experienced workers into the army, or by being attracted into industries associated with the war that paid higher wages. They worried whether they could retain any remaining men of military age [24] and feared that unless skilled labour was increased, they could not maintain productivity, let alone increase it [25].

By July 1915, government attempted to establish the numbers of both men and women aged 15–65 in the national workforce through the National Registration Act (how this worked is shown in a footnote in Chapter 2, p. 28). Results identified the number of men between 18 and 40 not in reserved occupations, their registration cards were handed to army recruiting authorities and committees to canvass them for army recruitment. The Act also identified the number of men and women with skills in key industries who should be retained at home [26–28]. Key workers designated indispensable to the war effort working in occupations of national importance were starred, identifying skilled farm workers,[5] while farmers and unskilled farm hands were omitted. When army recruitment did not increase, the Derby Scheme,[6] based on data

[4] The difference between reserved and protected occupations demonstrates the increasing rigour of providing men for the army by a system of call-up or exemption from military service. The definition is shown in Wolfe [20]. The police were never a reserved or protected occupation, the Chief Constable having the final say in who went to war and who stayed. This changed in 1918 when the police were given a quota to release.

[5] The main protected farm workers at this time were bailiffs, shepherds, stockmen including milkers, horsemen, machinery mechanics, steam plough and threshing machine workers (see Dewey [8]).

[6] The Derby Scheme secured pledges from all men aged 18–40 to join the army when called, by either joining immediately (around 11% did so) or deferring their army service and were then called up in sequence, the youngest single men first. However, without compulsion, only two-thirds of single men and half of the married men signed up, the remainder avoided recruitment [26]. For advice from the Board to farmers see Journal of the Board of Agriculture December 1915 Official Notices and Circulars pp. 903–5.

from the National Register was introduced. It disallowed starred workers from recruitment into the army, although they could attest their willingness to serve and were then placed on a reserve list to be called when needed, in the meantime remaining in their occupation.[7] Attested men could not be called without a military tribunal hearing, provided they remained in their starred occupation. Farmers were not starred, but the Board of Agriculture urged them to attest their willingness to serve but to defer their call-up, rather than signing up immediately. The necessary versatility of workers on small farms, often family members, frequently led to their being unable to claim exemption as only part of their work was protected, leading some farmers to give these workers work with animals to reclassify them as protected [30].

With the predicted failure of the Derby scheme and increasing pressure by the army that the war would continue for longer if more eligible men did not come forward, compulsion became inevitable. The first Military Service Act operated from 2nd March 1916 and was implemented in stages. It introduced conscription of single men aged 18–41 who were passed into the reserve, with a list of certified occupations replacing starring, to preserve manpower in essential industries which now included farmers, but still excluded ordinary farm labourers [31–33]. Consistent with Liberal policies of non-compulsion, the Act included application to a tribunal for exemption from conscription on certain grounds.[8] In the

[7] The starred occupations directly associated with agriculture were: engine attendant, driver and mechanic, bailiff, beastman, byreman, cattle- or cowman, dairyman, carter, foreman, horse-keeper, horseman, milker, milkman, ploughman, shepherd, stallion man, steward, stock-keeper, stockman, thatcher, wagoner and yardman. LGB Circular R.4 (15 December 1915) see McDermott [29].

[8] Grounds for exemption were:

> (a) … it is expedient in the national interests that he (the employee to be exempted) should, instead of being employed in military service, be engaged in other work in which he is habitually engaged or in which he wishes to be engaged or, if he is being educated or trained for any work, that he should continue to be so educated or trained; or
> (b) On the ground that serious hardship would ensue, if the man were called up for Army Service, owing to his exceptional; financial or business obligations or domestic position; or
> (c) On the ground of ill-health or infirmity; or
> (d) On the ground of conscientious objection to the undertaking of combatant service. (Military Service Act, 1916, pp. 4–5)

case of the police, the Chief Constable was responsible for applications to a tribunal. But exemptions caused conflict between different industries and with the army for the finite resource of fit men of military age, as industries resisted substitution and sought to retain their experienced staff as long as possible. General Macready, Adjutant General at the War Office from 21st February 1915, reported dwindling army recruitment and damaging strife between all government departments, each claiming theirs was the most vital to retain labour to win the war and all worked independently of the others [34, 35].

By the time government designated agriculture as a top priority due to decreased food imports, the labour force had dwindled considerably, although estimates of the numbers varied, the largest from the Board of Agriculture in October 1916 which used Board of Trade surveys[9] showed 350,000 in England and Wales had left either for war or for other industries. These estimated losses amounted to between 23% and 45% of the workforce since July 1914; monthly surveys showed the peak in January 1917 when the need for farm labour was greatest. However, responses from agriculture were consistently the lowest from any employer throughout the war, at only around 7%[10] [37], making the estimates conjecture; scepticism of their validity increased when the army acknowledged it did not keep records of recruits' previous occupations [38]. Furthermore, the surveys did not include small farms where family members often made a significant contribution [39].

Surveys also did not include the number and range of groups and individuals working as substitutes into agriculture who were increasingly important, they will now be considered in turn. Before 1875 farmers regularly used temporary, casual and seasonal labour at key periods such as ploughing and planting, and harvest, but pre-war this pattern became less frequent as they preferred to employ full-time staff [40]. Concern over one regular supply occurred from early 1915, when around 13,000 mainly male Irish migrant labourers who returned annually, 65% to England, particularly Cheshire and Lancashire [41] and Wales and 30% to Scotland, failed to return from 1916 due to the Easter Rising and fears of

[9] The Board of Trade conducted continuous sample surveys of labour losses from each industry from October 1914 quarterly throughout the war by surveying employers.

[10] For how these surveys were conducted and their inaccuracies see Dewey [36].

conscription, so were lost to British farming [42–44]. Farmers had relied on their help.

But alternative sources of manpower increased as farmers employed other previous casual labourers more permanently, in Scotland unemployed fishermen and miners entered agriculture [45]. School children gave a significant contribution, particularly at key periods, such as harvesting potatoes in England and Wales [46]. By January 1915, 34 of 62 local education authorities in England and Wales had given exemptions from school and by October 1916, 57 had released a total of 14,915 [47, 48]. Some authorities excused as many as 44% of their boys aged from 11 years from education by May 1916 [49]. Patchy evidence of release in Scotland [50] is less available. Schoolchildren were popular with farmers as cheap labour [51–53] but the National Agricultural Labourers' Union and George Edwards from Norfolk War Agricultural Executive Committee (WAEC) strongly protested at this exploitation of boys sometimes as young as 10, as did the Bishop of Oxford and some Farmers' Clubs [54]. Schoolchildren also returned during school holidays, some authorities altering the dates and duration to meet agricultural needs [55].

Two groups of women provided further sources of labour:

1. local village women likely known to farmers were not organised by central or local government making it difficult to assess their numbers, although they were likely a large and experienced workforce; Ernle claimed 210,000 were added to pre-war numbers [56]. Because their work was often by family members or informal local networks, the best estimates used trends in population data indicating a workforce of 80,000 full-time and 50,000 part-time in 1914, with a significant decrease by 1915, likely due to the war separation allowance to all women with husbands in the military. Full and part-time female workers increased in both 1916 and 1917 to exceed 1914 levels when the relative deterioration in the war separation allowance encouraged women to work; this link was not incidental [57]. They mainly worked part-time and seasonally, estimated as an equivalent of 80% of full-time, their contribution rose from an equivalent of 6,000 in 1916 to around 25,000 in 1917 [58] and 30,000 in 1918; Ernle estimated their total numbers were 320,000 in England and Wales by September 1918 [59, 60]. While there was considerable promotion of all women into agriculture and encouragement by the Women's War Agricultural Committees

(WWACs), the increased numbers can be attributed mainly to food price inflation.

2. From 1915 the Board of Agriculture attempted to increase the number and experience of women working in agriculture, mainly targeting educated women. A training scheme of 2–4 weeks established by the Board for women selected by Labour Exchanges placed 199 in farm work. Some local authorities also developed training schemes but catered for very small numbers. The county War Agricultural Committees (WACs) were also stimulated to organise women for farm work in autumn 1915, with a negligible result, followed by a circular from the Board in February 1916 ordering each WAC to form a Women's War Agricultural Committee (WWAC) to work with them and other local women's organisations [61, 62], but as the Board had no powers of compulsion at this stage, action was at best patchy.

By August 1916, 28,767 women were registered working in agriculture in England and Wales but many more did not register, making figures unreliable and included full-time and part-time workers and some who had worked on farms before the war [63]. Despite many private organisations also trying to stimulate women into agriculture, some provided training schemes duplicating the efforts of the WACs, little was achieved [64]. Much failure was attributed to the hostile attitude of farmers [65, 66]. From 1917 with the implementation of WAECs[11] with executive powers, further attempts were made to increase the number of women in agriculture.

A further source of previously untried temporary farm labour was established from 1915 when large numbers of soldiers from army camps were released on furlough for two weeks for the hay harvest [67], but their use in agriculture took many months to become established, due to differences in pay and working conditions. There is no indication of how they were chosen or if they possessed any agricultural skills or aptitudes, which made them an unknown quantity, unlike policemen. For autumn cultivation, furlough was increased to four weeks and from January 1916

[11] See pages XX below showing the development of interventionist policies for agriculture.

farmers could use them at any time of year for up to four weeks by application to the local Labour Exchange, although managers were instructed to only make requests in the absence of other unemployed labour. But soldiers were expensive compared with ordinary farm labourers and were paid on wet days, when it was customary not to pay agricultural labourers [68], making them unattractive. However, farmers' needs were evident as in 1916 the War Office agreed to release 27,000 from August, but many more were requested by farmers than were available: 12 August—13 October 28,805 were requested with 16,690 supplied. The shortfall caused bad feeling and distrust amongst farmers at the available level of army help, compounded by constraints on the hours soldiers could work [69–71]. Until the change of government in December 1916, soldier numbers released onto farms were relatively small and only temporary [72], although this drip-feed allowed government to assess whether they would be accepted or be challenged as government-imposed blacklegs paid at War Office wages, significantly higher than agriculture, and become the cause of industrial unrest.[12] Government was aware of the fierce opposition in other unionised industries to substitution of labour with the inevitable unrest and is developed below.

Taking account of all known sources of farm labour and estimating proportions of full-time work by women and children, Dewey re-evaluated labour shortages in agriculture from the start of the war until the end of 1916 finding reductions of only around 6% in 1915 and 9% in 1916 [73] (see Table 3.1), so that until late 1916 many farmers seemed able to fill vacancies [74]. Evidence from two English counties show the most losses were ordinary farm labourers [75]. However, by January 1916 labour shortages were the greatest problems for farmers [76], although losses from agriculture were somewhat lower than from other industries [77] due to four factors:

1. only 50% of the workforce was of military age, a lower percentage than other industries such as road transport [78];
2. from summer 1915 a modicum of protection was given to agricultural workers when army recruiting officers were instructed to

[12] A circular letter to WACs of 5th October 1916 from the Board of Agriculture and by the Local Government Board of 6th October 1916 showed that substitution of agricultural workers was already without significant protest TNA MH 47/142/4/3.

Table 3.1 Manpower working in agriculture in England and Wales 1915–1918 in '000s

	Pre-war	*1915*	*1916*	*1917*	*1918*
Conventional farm labour	1,543	1,429	1,377	1,337	1,347
Total replacements, including part-time equivalents of full-time work	Nil	15	30	74	135
Percent of pre-war total	100	93.58	91.19	91.45	96

Adapted from Dewey, 2014, Table 9.11 p. 140.

discourage enlistment unless replaced by women or men over military age, although this was often ignored [79, 80];

3. Dewey shows that low wages for agricultural labourers were unlikely to cause voluntary enlistment [81]; and
4. in smaller family-run farms, family pressure to remain may have dominated [82].

However, conventional farm labour of military age was far from secure, as the army continued to recruit from all industries, including agriculture. During March 1916 the deteriorating military position from huge loss of life in battle led to a review of certified occupations, concluding that if women and older men were used in key industries, more men of military age could be released into the army. On 25th March a revised list of certified occupations was issued by the Board of Agriculture to WACs: farmers full or part-time and indispensable were certified along with agricultural machinery, steam plough and threshing machine attendants, drivers and mechanics.[13] But certified status was removed from unmarried farm bailiffs, foremen, grieves and stewards under 31 years of age and unmarried stockmen, byremen, horsemen, wagoners, cattlemen, ploughmen and teamsters under the age of 25 and unmarried farm servants under 25 years of age in Scotland, all exemptions from military service for these men were revoked from 1st May. They or their employers were required to make a fresh application to a tribunal within 2 weeks for their exemption. The turmoil in calling up agricultural workers aged 18–30, combined with employers' applications to tribunals that they were essential workers and

[13] These occupations became certified as agricultural machinery was lying idle for want of manpower (see TNA MAF 36/58).

could not be released, led to despondency amongst farmers and to reports in the Journal of the Board of Agriculture of serious or very serious shortages of labour in every county of England and Wales [83–86].

The Board became sufficiently alarmed that it sent a circular to WACs in April 1916 suggesting they redistribute experienced labour between farms, from areas least affected to help farms where losses were so severe it was almost impossible to continue cultivation [87]. From the army's perspective, the many exemptions granted by tribunals, combined with recruitment pressures led to the revised Military Service (Session 2) Act of May 1916, extending conscription from single to married men aged 18–41 years, effective from 24th June [88].

1916 was a low point in agricultural labour, but the number of substitutes increased and with a slower decline in the conventional labour force, manpower stabilised into March 1918 at a loss of around 4% (see Table 3.1). Although farmers continued to claim labour shortages and unsatisfactory replacements throughout 1917, every effort was made to meet their demands as all available avenues were targeted. Farmers' claims of shortages were likely, particularly on small farms, due to replacements needing to be multi-skilled. Furthermore, as increased arable cultivation was more labour intensive than farming pastureland, merely replacing the numbers from before the war was insufficient to meet the demands for increased crop production [89]. Also, the physical strength of replacements by children and women made most unable to hand plough, the most pressing need in springtime 1917.

To provide more detailed information on the amount of labour and the type of agriculture undertaken by area, a survey of all farms over 5 acres and the labour force was developed by the Man-Power Distribution Board in August 1916 "to survey the whole field of possible supply and advise the Cabinet as to the most economical method of its employment" [90]. On November 7th, the Home Office wrote to Chief Constables in England and Wales on behalf of the Army Council, asking the police to distribute the survey forms to all farms over 5 acres and to collect the completed returns [91, 92], to discover "whether or not more men employed in agriculture in any district could be spared for the Army, or whether in any districts the available labour has already become unduly depleted." A similar request was sent in Scotland. Survey completion was required under the Defence of the Realm Act (Article 3), with penalties for non-completion. The letter from Sir Edward Troup, Permanent Under-Secretary of State at the Home Office, specified the inclusion of:

> (a) the land included in the holding, (b) live stock of all kinds, (c) workers now employed on the farm, including the farmer and all members of his family over 15 years of age, who reside on the farm, (d) casual labour employed, (e) male workers formerly employed who have left since August 4th, 1914 to join the Army. [93, 94]

The survey requirements were not only for livestock but also for permanent and casual workers employed since the outbreak of war. It provided national and local government and the army with a further updated source of essential data on national manpower[14] and animal stocks. Returns enumerated holdings over 5 acres in each county in England, Wales and Scotland. Delivering and collecting survey forms from each farm brought the police into closer contact with farmers.[15] Data were reviewed according to the Bath agreement[16] by the Commanding Officer and committees in each area, noting staffing as (a) sufficient (b) excessive or (c) deficient. Excess or deficiency was noted, e.g. ploughman, and "ominously" circulated to army recruiting officers [99, 100]. Findings showed some areas were well-staffed and others where it

[14] The results updated information from the National Register of 1915, see this chapter p. XX.

[15] Indeed, the police were asked on at least three further occasions to distribute survey forms or to place them in police stations for collection by farmers: (1) Board of Agriculture census of livestock in June 1916 [95], (2) census of livestock required by the War Office in April 1917 [96] and (3) Board of Agriculture agricultural returns for holdings of 20 acres or more in November 1917 [97].

[16] The Bath agreement gave the formula:

> The following scale has been agreed between the Army Council and the Board of Agriculture and Fisheries *as a general guide* for determining the number of men to be retained on farms, namely:-
>
> > One skilled able-bodied man or lad (wherever possible not of military age) for each of the following:-
> >
> > > each team of horses required to cultivate the land;
> > > every 20 cows in milk, when the assistance of women or boys is available;
> > > every 50 head of stall or yard stock, when auxiliary feeding stock is resorted to and the assistance of women or boys is available;
> > > every 200 sheep, exclusive of lambs, grazed on enclosed lands;
> > > every 800 sheep running on mountain or hill pasture. (italics supplied) [98]

was almost impossible to continue due to depletion. It was agreed not to call up further agricultural workers until further notice [101]. However, from the middle of April a deputation of WAECs meeting with Lloyd George agreed that women and soldiers would not be counted in farm labour according to the Bath formula [102].

Further help to retain agricultural manpower was by Lord Crawford's agreement with the Army Director of Recruiting, Auckland Geddes, to remove no more men from agriculture until 1st January 1917, or, if working in milk production, until 1st April, unless replaced by a soldier on furlough [103, 104]. But as army substitution officers visited each farm to identify suitable men and arranged with WACs for substitution, the pressure for release continued [105, 106].

National manpower struggles from January 1917

From early 1917 under the coalition government of Prime Minister Lloyd George, the struggle for labour between the army and essential industries at home intensified. Increased substitution into agriculture became crucial to expand crop production to avoid national food instability; government took control of agriculture in the national interest to supervise farmers. This era saw increased attempts at national manpower planning by establishing the Department of National Service, which initially failed to satisfy demands and was side-lined by major government departments that relentlessly competed for manpower.

Pressure by the army to sign up fit men of military age intensified from December 1916 following huge loss of life on the Somme and at Verdun and to overcome barriers by tribunal exemptions, the army called for release of 940,000 men between January and April: 100,000 during January, 170,000 in both February and March of which 350,000 category A men were needed[17] [107]. If these numbers were unobtainable,

[17] British Army Medical categories in 1914 were:

A. Able to march, see to shoot, hear well and stand active service conditions.

A1 Fit for dispatching overseas, as regards physical and mental health, and training
A2 As Al, except for training
A3 Returned Expeditionary Force men, ready except for physical condition
A4 Men under 19 who would be Al or A2 when aged 19

the army intimated fighting forces could not be kept up to strength [108]. 50,000 were identified from munitions, 30,000 from agriculture, 20,000 from coal mining, while Cabinet agreed to protect essential industries: steel production, agriculture, munitions, mines and quarries, railways, transport and shipyards [109]. The newly appointed President of the Board of Agriculture, Roland Prothero, protested that the country was increasingly frightened about food supplies and further release would hamper home food production, creating tensions between agriculture and the War Office, resulting in the appointment of Sir Arthur Lee as Director of Food Production to supply the needs of farmers through the WAECs. Lloyd George intervened on 22nd January ordering Field Marshall Robertson, Chief of Imperial General Staff, to send two divisions (between 10,000 and 50,000) from the Home Army to France[18] [110]. Subsequently the army agreed to enlist the 30,000 from agriculture slowly over 2–3 months while releasing 15,000 soldiers into agriculture to offset the withdrawal, each county would receive 250, with clerical help for applications from WAECs and farmers [111–115].

But the army remained pessimistic about achieving the remaining recruits, as by 2nd February only half of those required in January had signed up [116], so that the requirement for February remained 170,000.

B. Free from serious organic diseases, able to stand service on lines of communication in France, or in garrisons in the tropics.

- **B1** Able to march 5 miles, see to shoot with glasses, and hear well
- **B2** Able to walk 5 miles, see and hear sufficiently for ordinary purposes
- **B3** Only suitable for sedentary work

C. Free from serious organic diseases, able to stand service in garrisons at home.

- **C1** Able to march 5 miles, see to shoot with glasses, and hear well
- **C2** Able to walk 5 miles, see and hear sufficiently for ordinary purposes
- **C3** Only suitable for sedentary work

D. Unfit but could be fit within 6 months.

- **D1** Regular RA,RE, infantry in Command Depots
- **D2** Regular RA,RE, infantry in Regimental Depots
- **D3** Men in any depot or unit awaiting treatment

[18] The Home Army contained around 500,000 soldiers, anticipating the threat of invasion.

On 5th February the War Cabinet agreed to cancel all tribunal exemptions from men below 23 years of age, requiring their substitution by National Service volunteers, although recognising that substitutes would be less efficient than those released into the army [117]. The police throughout Britain contained large numbers of eligible men below the age of 23 (see Chapter 1).

The slower call up of 30,000 agricultural workers not engaged in milk production from 1st January was extended, agreed in Cabinet on 23rd January [118, 119] giving special attention to agriculture. On 24th January a joint letter from the War Office and the Board of Agriculture to County War Agricultural Executive Committees (WAECs) said no more men from agriculture would be called up unless replaced by substitutes unsuitable for military service; the army offered men graded C3—unaccustomed to farm work and only suitable for sedentary work. They formed agricultural companies and were available throughout the war. But by June 1917 only 9,800 companies had been established from the agreed 11,500, for example, Oxfordshire's was not set up until January 1918 [120, 121].

The army also agreed to return men with agricultural experience in steam ploughing when they could be traced and to send as many soldiers as possible to help with spring cultivation. This raised farmers' spirits that agriculture was being taken seriously [122]. From 28th February 1917, the Food Production Department was strengthened when more equal claims on manpower between army recruiting officers and WAECs were established, and farmers were advised how to obtain exemptions from military service through a tribunal application [123]. The police, as a non-certified occupation, similarly advised Chief Constables in February 1916 through their journal [124]. The role of Chief Constables in agreeing which policemen should be released into the army was confirmed in the 1917 list of certified occupations prepared by the Reserved Occupations Committee, issued by the Director of National Service:

> POLICE FORCES, MEMBERS OF – to be treated as in a certified occupation if their services are declared by their chief officer to be necessary in their civil employment. [125]

This gave policemen who remained at home an enviable position compared with colleagues as their work was deemed of national importance, increasing friction between individuals and claims of favouritism against senior grades as war progressed and the police were increasingly combed out.

However, the promised agricultural replacements were often not found before a key worker was called up, resulting in a deputation to the Prime Minister by the Federation of WAECs in England and Wales on 2 April 1917. As a result, an agreement was reached between the War Office and the Federation, sent to WAECs in a memorandum dated 5 May 1917. It overturned FP10 para 3 issued on 28th February on recruitment. Henceforth the WAEC's agricultural representative would notify the WAEC if a man (not the farmer or manager) was called up, depleting the farm's manpower unduly. WAECs should help to find substitutes and notify the Recruiting Officer to suspend the call-up notice for 3 weeks. With an indispensable farmer or manager, the WAEC should inform the Area Commander who, if he agreed, would cancel the call-up. Without agreement, the case was referred to the War Office and the notice suspended. Application for the withdrawal of a tribunal exemption would be after consultation with the WAEC or their representative. The Army Council proposed every call-up notice to a whole-time farm worker below category A should be certified by the employer. Each WAEC should appoint a tribunal representative to assist the agricultural representative, with appointments notified to the Area Commander and FPD. These measures increased the local balance of power between agriculture and the army so that "marauding recruitment officers" had less impact [126], protecting increased home food production for the 1917 harvest by providing some stability in manpower until March 1918.

The army released C3 men into agriculture as they were unlikely to make fighting troops despite army rations of 4,300 calories daily, above the recommended daily intake, to bulk them up [127]. The army wearied of feeding, clothing and billeting them, so sending them into agriculture provided a solution. In spring 1917 around 11,500 were sent to farms [128]; but farmers found most were physically unable to undertake the work and angrily rejected them, grumbling at being asked to keep infirmaries and insulted that they were to replace their skilled sturdy men. Farmers called them "lazy and useless" or "physically unable" and stated their conservative preference for men they knew and could rely on, or men who knew the district [129], and were not slow to return C3 substitutes to their depot [130]. In Scotland the difficulties were classified, and Scottish Command Orders gave appropriate actions:

DIFFICULTIES WITH MEN

1. How to deal with men who can work, but prefer not to do so. Directly a farmer returns a man in an Agricultural Company as being lazy and useless, the officer commanding company, if he is satisfied that the man is really only lazy, should transfer him immediately to the nearest Infantry Works Company or Works Battalion with a report.
2. If a man is returned because he is physically unable, he should be taken before a Board, recategorised to C (iii)., and then discharged as directed in Army Council Instruction No. 2333, of 14th December 1916.
3. Army Council Instruction No. 2333 was written before the Agricultural Companies were determined on, and therefore B (iii.) and C (iii.) men who were physically able to do agricultural work are available for Agricultural Companies, and should be posted in spite of this Army Instruction. [131]

The army had evident experience of those unable to undertake or who refused agricultural work to the farmers' satisfaction and had developed ways of dealing with them. Item 3 shows the types of men, only suitable for sedentary work, who were reclassified as suitable for Agricultural Companies.

From mid-March as the long harsh winter abated renewed urgency to plough and sow crops led the army to also release 12,500 Home Defence soldiers until 25th July, this was also when the first policemen were released for around 2 months to help farmers (see Chapters 6–8). The release of Home Defence soldiers was on their claims of previous agricultural experience. However, in early March 1917 when both C3 and Home Defence soldiers started to be dispatched to farms, only 3,000 (around ¼) were found to have ploughing skills [132]; many had made false claims to resist being sent to the front. Some were also highly unsuitable, including a lift attendant, a piano tuner, a needle maker, a chocolate maker and a ladies' tailor[19]; some also frustrated farmers at their lack of aptitude: "one of the men I got was a piano tuner. I could knock nothing into him" [133]; farmers ridiculed substitution [134, 135]. In response, on 12th March the army ordered all skilled ploughmen throughout Britain to return to their depot and allocated them to farms. By 10th April some 36,295 soldiers were in agriculture [136, 137] allocated until 25 July, but

[19] Agricultural Wages Board, 1919, p. 49.

temporary release was subject to army requirements making many of the fittest unreliable: 18,000 category A soldiers were recalled for operations in France before ploughing was complete. However, the remainder were allocated permanently along with the War Cabinet's decision in June not to conscript any further agricultural workers [138, 139]. But it was not until 1918 that the skills of army replacements improved, as their numbers increased [140], resulting in fewer complaints from farmers [141–144]. By 1918 permanent substitute soldiers willing to learn had become skilled workers [145]. But policemen released to work in their local area from March 1917 were selected by Chief Constables for their latent agricultural skills and likely to be known to farmers. Chapters 6–8 show the selection procedures and the locations of release of policemen. Most would integrate easily into the work and were highly regarded for their work on farms, while helping to supervise the array of substitute workers.

March 1917 saw further calls by the army with instructions to local tribunals to review urgently all exemptions for men in categories A and B1 under the age of 31 [146], this applied to the police but not to protected agricultural workers. Alongside increasing pressure on farmers to increase home food production, the effect of this further uncertainty was brought to the House of Commons by George Lambert, MP for South Molton, Devon:

> I have here three cases which appeared in the "Times" newspaper on 27th February. The first is a farm of 216 acres, 30 acres arable, 75 cows, 14 horses, 75 ewes, 2 store pigs, 7 miles from a station. The farmer had a son, who was rejected (by the military). He has now been re-examined and accepted. The farmer had three men, and if this young fellow is taken, as I understand he is to be, the farmer will be left with himself, who has a broken arm, and a daughter aged seventeen. ... the next case. A widow over seventy-two years of age farms 72 acres, 27 cows, 3 horses, 16 sheep. One man has been with her for seven years, and he does everything. This man has been taken. The military claimed him and offered a substitute, who turns out to be an artist and journalist. Here is another case: A widower, well over sixty, with chronic asthma and unable to fodder cattle, owns and farms 80 acres. He has one son, who does all the work, and no other labour. The military claimed him and refused a substitute...The War Office is entirely responsible. It is the uncertainty that is killing agricultural enterprise. No man can attempt to cultivate his land if he does not know what labour he is going to have. It has been taken away on every hand, and it is taken away very suddenly. [147]

These cases show in detail the drain on agricultural labour and how the military was able to overrule farmers' protests at labour shortages. Any substitutes particularly on small farms, around 292,000, 67% of all farms in England and Wales [148], needed a broad range of skills to compensate urgently for losses of experienced workers. Similar hardships were shown in Westmorland where elderly farmers appealed to tribunals for exemptions for their only son [149]. During March the Board of Agriculture continued to claim very deficient agricultural labour in England and Wales, eased somewhat by army substitution [150].

An additional 17,000 soldiers were released into agriculture for the hay harvest from June 1917 [151–153], along with more policemen (see Chapters 6–8) and another 33,000 soldiers from the first week in July, available permanently, unless replaced by men of similar quality [154]. But the 33,000 were late to arrive, only 26,000 were on farms by December 1917 when harvesting had been completed; and only around 2,500 had ploughing skills [155–157]. By March 1918 numbers rose to 63,000, most were C3 men from agricultural companies and by August 1918 to 70,000 with a further 10,000 to help with the harvest. In Scotland by late 1918, there were 9,000 C3 men with another 10,000 for harvest [158].

Other sources of farm labour also available in 1917 in addition to village women included on pages 44–45 above, were skilled or educated women intended as a mobile labour force, often from outside the local area, organised by the WWAECs [159]. They particularly responded to calls for patriotism [160]. The Board through the FPD established WWAECs in January 1917 to increase the supply of women on the land and were responsible for selection, training, equipment and allocation to farms including employment, housing and care while at work [161]. An appeal by the National Service Department was also for more village women to work part-time in their neighbourhood and to women in urban areas to join a full-time mobile Women's Land Army [162, 163]. But, although many came forward in 1917, their physical fitness was such that 75% were rejected [164, 165]. Numbers rose from 5,000 in July 1917 to 6,672 by November. Initially farmers doubted their help due to anti-female prejudice that farm work was degrading to women and feared their lower pay would threaten male employment [166] along with worries of providing accommodation for single women [167], due to fears of sexual immorality. Following a further recruitment drive in Spring 1918, numbers rose to 7,665 in March and 16,000 by September; the latter increase was partly due to each county strengthening and expanding welfare for the girls to promote their continued farm work.

From January 1917 female labour was additional to male labour, so could be placed without fear that their employment would automatically release a man into the army [168–171]. But they were not evenly spread across Britain with few in Wales and none in Scotland. They were eventually accepted by farmers who rated their work as efficient [172–174].

From April 1917 in the national campaign to plough, another source of help was by agreement between the Board of Agriculture and the Local Government Board to allow pensionable men in rural districts to earn wages without the loss of their pension, many men over 45 joined the agricultural labour force [175, 176]. Reports from Birkenhead showed "An interesting experiment has been in progress for some weeks near Birkenhead, where a number of gentlemen have organised themselves to help the local farmers. They have given the equivalent of 115 half days, and their work has been greatly appreciated" [177].

Temporary help for the 1917 harvest was also by around 5,000 public and secondary schoolboys aged from 15, numbers increased to over 15,000 for the 1918 harvest. WAECs arranged them into camps in England and Wales [178–181]. From 1917 the numbers of exempted elementary school children also increased, continuing into 1918 [182]. Although total numbers for Scotland are not readily available, Brown shows school boards exempted between 0.8% and 7% of schoolboys aged from 12 years of age during 1914–1918 from Forfarshire, Perthshire, Fife and Dundee [183]. Children also worked on farms during school holidays.

Other substitutes were prisoners of war (POWs), under the control of the army. By the war's end they were the second largest group deployed to farmers after soldiers [184], but placement on farms was slow. Although in January 1917 WAECs in Cheshire, Hertfordshire and West Sussex applied for batches of 75 with 35 soldier guards offered by the Board of Agriculture, both Oxford and Wiltshire said these numbers were excessive. In February the War Office agreed to reduce batches to 40, Somerset requested four batches and Norfolk expected 150. But by August 1917 only 15 camps were established housing 1,476 [185]. Not only was accommodation for over 40 men a problem in many areas, but they were not popular locally and so confined to the new land drainage schemes rather than working in farming [186]. With new targets for increased arable acreage communicated to WAECs in June 1917, ploughmen continued to be in short supply and were found amongst POWs but ploughing required them to be spread over wide areas, such as small farms with scattered fields in Westmorland [187], causing problems of how to provide sufficient guards reflecting worries about escape and hostility by local residents, which gradually became unfounded. Relaxation of guarding was agreed in the War Cabinet on 13th December 1917 allowing additional POWs with ploughing experience to be recalled from France [188–190]. By January 1918 ploughing camps of 30–40 were agreed accompanied by guards at night and by either a soldier or a policeman or ploughman by day, acting as foremen. Local scepticism had sufficiently diminished by February that up to three could be boarded on each farm [191]. This heralded the growth of numbers in farming; February 1918 saw 4,279 out of which 692 were in land reclamation, market gardening and land drainage [192]. They were joined shortly afterwards by a further 4,713 of whom 3,771 were ploughmen, very welcome at a critical time of year [193]. By May numbers were 9,783, and by October a dramatic increase to 25,213 employed in agriculture [194, 195]. They were generally inefficient and slow, although conscientious, mainly attributed to inadequate feeding until July when farmers were obliged to provide a midday meal [196].

Providing sufficient labour for the 1917 plough policy was a continual struggle, consistent with all industries supporting work of national importance [197–199], farmers continued to complain of lack of trained ploughmen, particularly in the crisis of food supply in early 1917 and into February 1918 with monthly reports from each county in England and Wales published in the Board's journal showing that the shortages were patchy; some areas suffered more than others. The inexperience of many temporary workers also put a heavy burden on farmers [200]. Meanwhile, government exerted tight control of agriculture while seeking to appease farmers to increase crop production.

This chapter so far has shown the large number of different groups and individuals lent to farmers from around 1916, most were temporary and many unskilled in agriculture. Farmers preferred to work with people they knew and trusted although most of the substituted labour fitted neither of these categories when they were initially allocated to farms. With pressure on farmers to meet government targets for crop production, as the next section shows, farmers welcomed police help, not only for physical work, but also for the surveillance they provided over this unknown workforce, although no instances were found of prosecution of substitutes for stealing produce or misbehaving.

The next section demonstrates how the revised constitution and workings of the WACs under the Board of Agriculture from January 1917 gave them responsibility for supplying substitute manpower to farmers and how they persuaded farmers to comply.

State control of agriculture to feed the population 1917–1918

Wide-ranging interventionist policies were speedily developed from December 1916, when Lloyd George became Prime Minister, to increase home food production and to reassure the population that everything was being done to avert impending national starvation, and the threat is shown in Chapter 2. Every avenue was explored to help farmers expand arable production, the most important was an increase in competent labour when much tilling was by hand ploughs pulled by horses. Government policies were two-fold:

1. The appointment of Roland Prothero as President of the Board of Agriculture on 13th December 1916[20] followed by establishing the Food Production Department (FPD) in January 1917 intended to boost the needs of agriculture against manpower claims by the War Office and heading the system of county War Agricultural Executive Committees. Prothero described policies as "the improvement and extension of arable cultivation, with spade as well as plough; decentralisation; and drastic powers of compulsion which could only be justified or tolerated in a war emergency," intended to turn back the clock to before the decline of agriculture in the 1880s [201, 202], although he recognised the nation had little interest in agriculture, except when food supply became limited [203]; and

[20] https://www.thegazette.co.uk/London/issue/29865/page/12225 (accessed 7/3/2024).

2. By establishing the Department of National Service headed by Neville Chamberlain also in December 1916, with the intention of balancing the competing manpower demands of the military and essential industries at home [204, 205].

Prothero's 20th December 1916 speech to the Federation of War Agricultural Committees outlined the policy of:

1. increasing arable acreage and fixing prices, inspired by the Interim Report of the Milner Committee of June 1915 (See Chapter 2, p. 27);
2. compulsory powers; and
3. decentralisation from the Board to local authorities.

The "plough policy" was to increase output on existing arable land and to bring derelict and under-used land back into full cultivation [206, 207]. His speech was followed by circulars to WACs on 28th December giving the aims for farming, and 29th December specifying the methods. He entreated farmers to respond in the national interest:

> We have two objects before us. One is immediate – the spring sowing in 1917. The other is more remote – the preparation of an arable area for the next cereal year. Farmers want – and the nation expects of them – to grow all the essential food they possibly can, both for human consumption and for the maintenance of live stock. [208]

The circular estimates the war continuing beyond 1917 and entreats farmers to step up in the national emergency to provide for the nation while flattering them that this is what they also want and encouraging them to accept that it requires nationalisation. The first circular identified acreage under each crop in England and Wales and the proportions of wheat, barley, oats, beans and peas compared with imports, showing only one fifth of wheat was home-grown, stressing the "urgent necessity" for increases. Farmers were implored to plan their crops for a harvest to feed the nation [209]. The second letter gave 4 aims: (1) Organisation of WACs; (2) Surveying land; (3) Ploughing grass; (4) Organising potato growing and pig-keeping [210]. Advice on a land survey used Essex WAC's example [211]. The most contentious was ploughing grassland to increase tillage, some agriculturalists believed all available manpower should be devoted to existing cultivation rather than to increases [212].

The plough policy used scientific evidence from the Food (War) Committee established within the Royal Society in 1916 [213] to show how the change of land use would help to feed the population:

100 acres of very poor grass fed to cattle	could feed 2–3 people for 1 year
100 acres of medium grass fed to cattle	could feed 12–14 people for 1 year
100 acres of very good grass fed to cattle	could feed 25–40 people for 1 year
100 acres of fodder beet (mangolds)	could feed 35 people for 1 year
100 acres of wheat as bread	could feed 200 people for 1 year
100 acres of potatoes	could feed 400 people for 1 year[a]

[a]These values were allegedly adapted from the Committee's findings [195]

The policy of increasing arable production while decreasing pasture-land and grazing had its critics as farmers were unable to implement it without increased labour, horses, fertilisers, seed, supplies and equipment [214–216]. The above quotation of land usage can also be seen not only to feed more people per acre, but to cater more for the dietary habits of the majority of the population, as I set out in Chapter 5.

Decentralisation from the Board to WACs required reconstitution from the advisory bodies established under Lord Selborne[21] (see Chapter 2) into executive committees charged with carrying out the Board's directives. The 23rd of January notice to WACs gave their constitution and powers to direct cultivation [217]: They must restrict their previous unwieldy membership to between four and seven with a District Commissioner to cover two to three counties representing the FPD [218–220]. Most achieved the transition into county War Agricultural Executive Committees (WAECs) by January comprising farmers, landowners and others interested in agriculture; Norfolk was unusual in appointing an agricultural trade union representative, George Edwards [221]. However, some counties resisted, as all their districts were not represented as we see in Cheshire in Chapter 6; a request to the Board for an increased membership was declined [222]. WACs in Scotland were similarly formed into Local Agricultural Committees of between five and ten members in each local government district [223]. But the Scottish Board of Agriculture retained responsibility for land surveys and issuing compulsory

[21] For Lord Selborne's contribution to agriculture in the first two years of the Great War, see Chapter 2.

ploughing orders [224] and for food production [225]. The distribution of substitute soldiers and others into agriculture in Scotland was also by the Board of Agriculture, while the Department of National Service Agricultural Section located the labour [226]. How these structures helped farmers at county and district levels and how the police were involved is detailed in Chapters 6–8.

The Board created the FPD in England and Wales from 1st January as a separate department with independent day-to-day working heading the 61 WAECs [227–229]. The Director-General from 19th February was Conservative MP Arthur Lee, responsible for providing labour, often from military depots in each county, mechanical supplies such as tractors and spare parts, coal for steam ploughs, horses, fertilisers and supplies such as seeds; although substitute soldiers and horses ultimately depended on the War Cabinet; the Ministry of Food retained responsibility for horse feed; while the machinery, implements and fertiliser manufacture was with the Ministry of Munitions; the FPD was responsible for sourcing and distribution to farmers [230, 231]. Each WAEC provided a weekly report to the Board showing actions under Defence of the Realm (DORA) Regulations 2L and 2M[22] [232].

The strengthened remit of the WAECs and formation of the FPD in England and Wales provided clarification and more vigorous defence of agriculture and its manpower, despite continual external pressures to release men into the army, seen in the above sections of this chapter. Many farmers were reluctant to accept substitutes and refused to pay increased wages [233]. However, they were persuaded to increase arable cultivation with guaranteed prices for growing wheat, oats and potatoes from the 1917 harvest [234–236], DORA Regulations 2L and 2M [237] also enforced cultivation in England, Wales and Scotland. DORA principles were later enshrined in the Corn Production Act on 21st August 1917[23] [238].

WAECs appointed full-time salaried Executive Officers knowledgeable in the local area who accepted a heavy workload. Initially WAECs speedily

[22] For the exact wording see https://archive.org/details/realmdefense00grearich/mode/2up?ref=ol&view=theater (accessed 15/1/2024) pp. 58–65, also Middleton pp. 349–362.

[23] Except Part 4 of the Corn Production Act which allowed farmers who disagreed with the WAECs proposals to go to arbitration, which was not implemented until August 1918.

surveyed the land with local surveyors often accompanied by farmers; they graded farms as: (1) well farmed; (2) indifferently farmed but capable of improvement; and (3) derelict. Surveys were delegated to District Committees, where these existed, usually based on rural district council areas [239]. The survey was speedily followed by WAECs issuing thousands of ploughing orders to farmers with the right of appeal to the WAEC if they disagreed [240], however, only around a quarter of farms that had the potential to increase arable production received an order. Many WAECs invited dissenting farmers to their meetings to persuade them of a course of action. Only if persuasion failed with repeated attempts to encourage cultivation was prosecution pursued [241], farmers could be fined or imprisoned for failure to comply [242]. However, litigation was used sparingly, only 251 from 100,000 orders resulted in litigation [243–245], although cases were widely publicised in the press. Decisions to enter the land for cultivation or to dispossess the owner were referred to the Board, giving considerable work to the WAECs [246]. The Board's options were: landlords should terminate the existing tenancy and replace tenant(s), or the land could be taken over and cultivated by an appointee of the WAEC, often a local farmer [247, 248]. WAECs could also take direct responsibility for the land, buildings and equipment, as in Essex and Kent [249] but this could be costly requiring the local authority to increase land productivity indefinitely, an unattractive option if produce and/or land prices fell [250]. WAECs also negotiated to overcome covenants forbidding tenants from ploughing their pasture [251–254[24]]. Under DORA, farming became a nationalised industry. By 1918 finance and drainage sub-committees were added to the WAECs workload.

Farm workers were encouraged to remain on the land by a minimum wage of 25s per week from 21st February (a wage also offered to National Service Volunteers), in the hope of tempting former agricultural workers to return to the land, but was voluntary until between May and October 1918 when implemented by the Agricultural Wages Board [255], while steps were taken to prevent increased farm rents, benefitting

[24] The journal showed how WAECs could inspect the land at the tenant's request, if they agreed it could be broken up, WAECs could issue a direction to the tenant to protect him from any proceedings taken against him by the landowner. Landowners expecting a loss could claim compensation under DORA (Losses) Commission.

tenant farmers and farm workers in tithe cottages. However, the potential for farmers to exploit farm labourers by expecting an increase in their hours by one hour per day would be "a death-blow to allotment cultivation" as farm labourers were currently producing food for themselves and a surplus for others on their allotments, the Corn Production Bill should prohibit farmers from increasing labourers' hours to compensate for increased wages [256]. The success of these measures was demonstrated by the Agricultural Wages Board of the Board of Agriculture in 1919 showing farm workers' rents increased by 8.7% from pre-war, whereas food and fuel prices escalated to between 83% and over 100% [257].

Under the Board's letter of 23rd January [258] WAECs were to appoint District Committees to work locally, entrusted with assisting and encouraging farmers to cultivate their existing arable land for the greatest output of essential crops and by surveying land to identify where pasture could be converted more profitably into arable, requiring it to be broken up [259]. They were the lynchpin of assessing farmers' needs responsible for reporting to their local WAEC. But District Committees were sometimes slow to deliver and were reprimanded and encouraged by Ailwyn Fellowes,[25] Chairman of the Norfolk WAEC, appointed vice-chair of the advisory committee on increased home food production:

> AGRICULTURE AT WAR.
>
> Sir Ailwyn Fellowes presided at a meeting of the chairmen and secretaries of the Norfolk District War Agricultural Committees at Norwich on Saturday, when he had a straight talk with them. He directed his remarks to the urgent importance of the District Committees doing even more effective and useful work than hitherto. ...
>
> Sir Ailwyn reminded the meeting that fifteen months ago the late Government appointed War Agricultural Committees in every county, and the first thing they did was to see that district sub-committees were appointed. He had hoped that the district committees would have given the County Committee great assistance in the work, ... but with the exception of a few committees it had not been as satisfactory as could be wished. It should be realised that the work of the district sub-committees was the life-blood of the central committee, because they knew the wants and interests of the districts. The time had come to speak quite plainly, and for him

[25] Ailwyn Fellowes was Minister for Agriculture from March to December 1905, when the government fell.

> to say that it was on the district committees that in future would depend the way farmers and others in Norfolk were to receive assistance and help from the Government through the County War Agricultural Committee. He understood there were certain members of district committees who never took the trouble to attend the meetings. All he could say was if these people were not patriotic enough to help their brother agriculturalists in this crisis then the sooner they left the Committee the better. If the members would not meet the best thing was to ask them to resign and let men who would undertake the work fill their places, because duties of the Committees in future would be heavy and strenuous. [260]

Ailwyn Fellowes made this strong appeal to local patriotism and also to feelings of brotherhood to increase local crop production. His appeal was to local farmers to exert influence locally by becoming involved in farming for the national effort during the war and to make a contribution accordingly, rather than to be absorbed only in their own farms. The districts' crucial role was to provide the county committee with detailed information on which they could apply for assistance for farmers.

In May 1917, anticipating more arable land for the 1918 harvest, the FPD also requested WAECs to establish District Committees. Many were efficient but some needed allocated work to meet regularly. If they continued to be inefficient the WAEC could establish an enquiry and supersede them if they did not improve, including better co-ordination between the men's and women's sections [261]. As the districts' work increased, the FPD allowed appointment of a District Executive Officer and devolved many of the WAECs powers; they were advised to subdivide into smaller areas such as parishes to manage the plough policy more locally, establishing intense surveillance of farming [262, 263].

However, despite strengthening agricultural politics locally, the army's call for manpower continued. This precipitated a national approach to manpower planning to protect priority industries while providing for the fighting forces.

National manpower planning and redistribution

The army's call for 940,000 men in late 1916 (see p. 50 above) and their recommendations for compulsory National Service for men between 18 and 41 [264, 265] sparked a national manpower crisis. By December 1916 the Asquith government's failure to organise civilian redistribution, particularly criticised by Lloyd George, and the Man-Power Distribution

Board's frustration that their report and recommendations of October had been side-lined [266], contributed to Asquith's downfall as Prime Minister on 5th December and the appointment of Lloyd George [267].

From 7th December Lloyd George, vociferous in the need for national manpower planning, gained agreement for a Department of National Service to enrol men and women 18–61 years of age voluntarily, if volunteering was unsuccessful civilian compulsion would ensue [268, 269]. Enshrined in the department, replacing the Man-Power Distribution Board, was mobilisation of the civilian population and to balance the supply of workers between essential industries and the military, to help protected occupations while minimising labour in work not directly contributing to the war effort, such as building, printing, paper, hospitality, entertainment, commerce and professional occupations [270]. These workers should be transferred as substitutes into essential industries for more men to be released into the army, an estimated 1,000,000 could be transferred, responding to the army's intense pressure for 940,000 (see p. 50 above). Lloyd George told parliament:

> … the illustration of agriculture. There is undoubtedly in this country a considerable number of people skilled in tillage of the soil who are not producing food, but we cannot mobilise them. We cannot direct them. I believe that there are scores, if not hundreds of thousands of people of that kind—there is no question here of military age—who, if we could utilise them to the best advantage, could produce great quantities of food in this country, but we cannot do it. Not only that. The difficulty in agriculture is the want of skilled men. You may have two or three skilled men on a particular farm, or the farmers may have no skilled men at all, yet two or three skilled men, if you could treat them as commissioned officers, could look after not merely one farm, but several farms, with the aid of unskilled men or women working under them. [271]

Lloyd George's speech came within two weeks of his appointment as Prime Minister. The need for national redirection of labour was clear; with the movement of skilled people who possessed recent or current abilities from industries not directly supporting the war effort into essential industries such as agriculture, this would release more men of military age into the army. This was to become the remit of National Service, initially based on voluntary recruitment. Neville Chamberlain reluctantly accepted the post of Director-General of National Service on 19th December

1916 from his post as Lord Mayor of Birmingham, at a highly controversial time in national manpower negotiations. A little known national political figure and not the first choice by many in government, his credibility with major government departments was low [272]. Adjutant General Macready, in control of the War Office from 21st February 1915, favoured Brigadier-General Auckland Geddes who he said sufficiently understood the extreme complications of manpower in Britain at war having built the army's recruiting system, recognised the inexperience of Chamberlain in Whitehall politics [273].

Most government departments feared the effects of National Service on their workforce, particularly the Ministry of Munitions and the Board of Agriculture [274]. Lloyd George advised Prothero in mid-January not to rely on substitutes from National Service and to request the Home Army's help with men unfit for general service and to enrol other volunteers for farm work [275]. The scramble by each major department to retain its skilled workers and for manpower generally during December to April was highly contentious, causing strife between government departments.

National Service was launched in Central Hall, Westminster on 6 February 1917. The campaign appealed to patriotism, while minimising compulsion everyone should place their labour at the disposal of the State, portrayed as restoring the nation's fortunes in this third year of war, to counter the German submarine campaign and to increase land under cultivation [276, 277], although the meeting was described by *The Times* as "rather undemonstrative" [278]. Local authorities were implored to help by establishing National Service Committees in towns and districts to organise publicity by public meetings to promote enrolment. Wages for volunteers were set at the current district rate for each trade with a minimum of 25s and travel and subsistence if working away from home. Chamberlain initially claimed he expected 500,000 volunteers during an 8-week publicity campaign. However, Middleton for the FPD said:

> In February 1917 the Ministry of National Service made an effort to secure volunteers for agricultural labour from among workers in less important industries. A number of men came forward, but relatively few were suited for farm work, and no important aid was got from these volunteers. [279]

Despite this dismal result, as Chapters 6–8 show, release of policemen across Britain was not reported to the FPD, showing the inadequacy of government manpower statistics in early 1917; the FPD was unaware of police help until 1918, when the reconstituted Ministry of National Service under Auckland Geddes included them.

National Service was controversial from the outset. Prothero retained responsibility for agricultural manpower and substitution in England and Wales, while Chamberlain maintained responsibility in Scotland, these differences in 1917 are shown in Chapters 6–8. Between 17th February and 1st March Chamberlain spoke at large public meetings in 5 cities: Bristol, London, Glasgow, Sheffield and Cardiff to promote National Service volunteering. He focussed on one priority industry in each location, in Glasgow his appeal was particularly for agriculture (see Chapter 8). However, the small number of meetings organised by National Service Committees combined with difficulties of door-to-door canvassing by voluntary workers to enrol volunteers, frustrated Lloyd George [280].

Recruitment in February 1917 received many applications, but few were suitable for agriculture [281]. A further appeal in March was also unsuccessful [282], although Chapters 6–8 show substitutes by policemen from March 1917. By the end of March, with only 388 men placed in all industries, it was clear government had completely misjudged the remaining goodwill of the population to enrol in a voluntary mobile reserve of labour. The dismal level of placement in notified vacant posts continued, despite the demands for labour and any successes Chamberlain achieved were undermined. Personally, he also lacked tact by showing competition between English and Scottish agricultural substitution in a rather boastful speech in Manchester on 27th March:

> A number of their (National Service) volunteers were placed and were doing a good service to the State. He was able to say there was no farmer in Scotland to-day who stood in want of a ploughman.

His claim of success in alleviating manpower problems for farmers in Scotland received wide publicity [283] when English farmers were still complaining of ploughmen shortages. The claims were confirmed by the Secretary of State for Scotland in the House of Commons on 2nd April in a challenge by Cecil Harmsworth, a member of Lloyd George's secretariat [284]. National Service Week (24th–30th March 1917) was intended to boost lagging recruitment and showed that agriculture had benefitted from between 500 and 600 skilled men lent by corporations and private employers for six weeks [285]. However, while National Service was based on voluntary recruitment, little impact was made on unionised areas of work, such as the priority industries of engineering, shipbuilding and mining.

After 5 months, Chamberlain's disillusionment is seen in his report of June 1917:

> Further reports confirm the impression derived from Reports made by Substitution Officers that up to the present there has been no demand for substitutes, for the reason that no men are being released from Munition Works for Military Service. Such volunteers as have been placed have been by way of reinforcement and not of substitution. [286]

This confirmed the rationale to provide voluntary substitutes for the release of fit men of military age had not been achieved, substitutes had merely been absorbed into the workforce without the compensatory release of men of military age into the fighting forces. The weakness of National Service was further shown by Macready's refusal to assign army recruitment, as the Department was unable to assert its authority over other government departments [287]. Chamberlain resigned in August and returned to Birmingham, to begin a life-long rift with Lloyd George, who he claimed did not support him. His help for agriculture in Spring 1917 has not been made public to date, although this book shows differently (see Chapters 6–8).

National Service as a voluntary scheme raised fears of industrial conscription and strike breaking in major industries and was fiercely contested by Labour and the trade unions, particularly in engineering and munitions, setting up barriers to deploying substitute workers from outside the industries [288–291]. However, agriculture was largely non-unionised so mainly outside these hostilities, although strikes by farm labourers occurred in Berwickshire and part of Roxburghshire for higher wages during passage of the National Service Bill [292] (see Chapter 7). The imposition by government of huge supplies of labour into agriculture, seen by agricultural labourers as government-assisted blacklegs, weakened their ability to bargain for wage rises in this low wage industry, until the establishment of the Agricultural Wages Board in 1918. Additionally, many substitutes were exempted from conscription, while agricultural labourers were being called up [293–295]. Until March 1918 unskilled farm workers received lower wages than substitute soldiers, paid by the War Office [296] and friction between farmers and labourers was admonished since at least the end of February 1917 [297]. But few reports were found of strike action in farming [298] leaving government with little opposition to deploy substitutes; and while farmers received a minimum price for their produce as inducement to increase production and were supplied with essential manpower and equipment, there was little resistance to imported labour. Whereas farm labourers were exploited by government, the police received their existing wages and a subsistence allowance if working away from home. In some areas the increase paid by this system allowed the police to give a bonus to members of their force (see Chapters 6–7).

National Service was reconstituted based on compulsion in September 1917 under Auckland Geddes, previously Director of Recruiting at the War Office. The Department, raised to a Ministry included a seat in the House of Commons, increasing the profile of national manpower planning [299]. Geddes' record at the War Office gave him credibility in government. Unlike Chamberlain, he agreed a set of "defining functions" for the Ministry:

1. To review the whole field of British man-power and to be in a position at all times to lay before the War Cabinet information as to the meaning, in terms of man-power and consequential results, of all Departmental proposals put forward to the War Cabinet, and referred to the Ministry for its consideration and for an expression of its opinion.
2. To make arrangements for the transfer from civil work, not declared by the War Cabinet to be of primary importance, or if ordered by the War Cabinet from the Navy, Army or Air Service to urgent national work, or such numbers of men as may be declared by the War Cabinet to be necessary to reinforce the labour already engaged on that work.
3. Subject to the approval of the War Cabinet to determine, in consultation with the Departments concerned, the relative importance of the various forms of civilian work, and to prepare from time to time lists of reserved occupations, with such age and other limitations as may be necessary to secure the maintenance of essential public services, and the preservation of a nucleus of civil occupations and industries.
4. Within numerical limits imposed by the War Cabinet to obtain for the Army, Navy and Air Service such men as can be withdrawn from civilian life without detriment to the maintenance of essential public

services and the due performance of the civil work necessary to maintain the forces at sea, in the field, and in the air, and any nucleus of civil occupations and industries declared by the War Cabinet to be necessary.
5. In connection with Function 4, to determine the physical fitness of men available or possibly becoming available for withdrawal from civil life.
6. To make arrangements for the provision, where necessary, of labour (male or female) in substitution for that withdrawn from civil life in accordance with Function 4. [300]

By stating the defining functions of the ministry based on compulsory movement into priority industries and agreement with the War Cabinet and other government departments, Geddes immediately strengthened his position as able to determine and control manpower throughout Britain. His credibility with essential industries with strong unionisation, such as shipbuilding and mining, who put up barriers to substitution and protected their skills from dilution by unskilled workers, was particularly important. Unlike National Service under Chamberlain, work of national importance became compulsory for everyone between 18 and 61, combining army recruitment with the movement of labour into vital war work [301]. It became central to manpower review across all government departments linked to strategic priorities and the final arbiter in national manpower from November 1917 [302]. The Ministry also recorded substitute and part-time labour, such as policemen. Thus began a more settled period for manpower with a limit on the drain of skilled and experienced men away from industries supporting the war effort, until early 1918 [303, 304]. The FPD depended on it for the supply of agricultural labour [305].

In October 1917 Geddes assessed the army's strength by Category A men and anticipated the numbers to maintain army strength to October 1918. He concluded those in civilian life were already engaged in essential industries, while the remainder were in poor physical health and not suitable for active army service. If the conscription age was raised to 51, 270,000–280,000 Category A men working in non-essential industries would become available [306]. In December 1917 a Cabinet Committee on Manpower agreed to review all military exemptions on occupational grounds, enacted in the Military Service Act 1918. If needed, the age limit would be raised to 51 [307].

Geddes' speech in the House of Commons in January 1918 showed a turning point in more equal claims between industries at home supplying the war effort and feeding the population, currently with 420,000 vacancies, alongside calls by the army for more troops to counteract Russia's impending secession with movement of enemy troops to the Western Front:

> National Service is not a question only of raising men to fight—it is a question of maintaining an even poise and a balance between all the activities of the State, all those activities which are essential to the winning of the War. [308]

While Geddes was not sympathetic to tribunals, indeed, by early 1918 he regarded their extinction as desirable, his aim was to ensure the maintenance of war production, particularly munitions, food and shipping, ranked at least equal with the Army's need for men [309].

Alongside national manpower planning plans to increase home food production continued into 1918.

Continuing the Plough Policy into 1918/19

Despite the difficulties of agricultural manpower and supplies (the supply of horses and mechanical ploughing equipment forms Chapter 4), government persisted with the plough policy into 1918 and attention turned to the harvest. The Board proposed an increase of three million acres to be ploughed in addition to the 1916/17 acreage. An editorial in the Board's journal enforced the need for all agricultural land to be devoted to food production, framed as patriotism "It must be clearly understood that henceforth bad farming is a danger to the State, and that the waste of good land on game or games is inconsistent with patriotism" citing the maintenance of golf courses as unpatriotic. The Bath formula for manpower was re-introduced (see p. 49), stating:

> It is obvious that this scale cannot be undeviatingly followed, owing to the variety of methods under which agriculture is practiced in different parts of the country, *nor does it embrace all descriptions of men employed on the land;* it is circulated merely with the intention of affording as to the *minimum number of certain classes of men* who are required to *preserve* the farming industry. (italics supplied) [310]

This quotation reminded farmers of how manpower will be assessed if a request to retain male labour of military age on a farm is taken to a tribunal. It is particularly timely as the targets for a 1918 harvest were delivered to farmers in June.

On 14th June 1917 Arthur Lee notified WAECs of the requirements for the 1918 harvest [311]. WAECs asked District Committees to conduct a more detailed re-survey of land [312]. Cultivation targets were downgraded from 3 million acres to 2.6 million acres in England and Wales, and 0.35 million acres in Scotland. The proportion for each county was sent to WAECs on 14th June, [313–315] for Cheshire this was an increase of 33% over 1916 [316, 317]. The War Cabinet recognised farmers would need vast increases in labour, equipment and supplies to meet these targets with the bulk of substituted labour from the army, agreed on 27th June 1917 as 25,000 experienced men and 25,000 unskilled men, provided at 5,000 per week which started 7th July [318]. However, pressure on the army allowed release of only 30,000 by December and the full 50,000 did not arrive until March 1918 [319, 320]. Agricultural manpower was debated in the House of Commons in January [321] followed by requests to men "of adequate means" to help on farms, by mid-March 9,650 skilled ploughmen had responded, well below farmers' requests [322].

By 21st March 1918 the German Spring Offensive on the Western Front caused a renewed army emergency, as large numbers of enemy troops from the Eastern Front with Russia were sent to fight on the Western Front. The offensive was timed before most American soldiers arrived in France. Britain responded urgently in the Military Service Act (No. 2) [323] raising the military age to 51 and called for more men. Exemption certificates were removed from farm workers aged 18–23 on April 20th [324, 325] and agriculture ceased to be protected. Board of Trade estimates for 1917 and 1918 showed 177,000 men of military age still working on farms [326]. Although Prothero had proposed an additional 3,400 skilled men by autumn 1918 for the 1919 plough programme [327, 328], the Board agreed with the War Office on 8th May that no more than 30,000 between 19 and 31 should be released in England and Wales and 5,500 from Scotland. The FPD reluctantly gave each WAEC a quota, asking for men whose release would cause the least damage, which many found difficult to meet [329, 330]. The demands caused public uproar at potential loss of the 1918 harvest, with deputations from WAECs and questions in both Houses of Parliament, the House of Lords on behalf of the National Agricultural Council comprising many landowners was particularly critical of National Service releasing more men from agriculture [331]. While 15,588 between 19 and 23 were called up, only 7,000 aged 24–31 were released before 27th June as their skills were difficult to replace [332]. However, Middleton said the effect on the 1918 harvest "was not considerable" [333], despite between February and April one of the largest acres of grain in living memory had been sown, which needed significantly more labour and equipment to harvest [334]. The land ploughed and the harvests produced are shown in Chapter 9.

Severe criticisms of WAECs combined with labour and supply difficulties led Arthur Lee to resign as Director of the FPD in protest, as any plough programme for 1919 had become a political liability and was abandoned. WAECs, however, ploughed and planted arable land in spring 1918 with around 60,000 substitute soldiers and more supplied for the harvest; by the end of the war some 84,000 were on the land. Soldiers were the largest single source of labour supplied to agriculture [335–338].

In June 1918 the arrival of American troops on the Western Front [339] eased manpower pressures and anticipated the end of the war. After harvest 1918 no more men were needed by the army, shipping had been protected by convoys since 10 May 1917 [340] and food imports stabilised [341], shipbuilding and repair started to increase, and food rationing was established. However, food supply did not increase, as shipping was needed to transport troops, equipment and supplies from America, reducing the space for food imports; the Allies were told to make supreme sacrifices with no shipping available [342, 343].

Conclusions

This chapter started by giving the large number of groups and individuals who provided temporary help for farmers in the crisis of increased home food production to feed the nation, and the largest group was soldiers who initially were of variable quality and reliability. At national level the struggle developed into the huge and increasing competition for manpower, where everyone from 16 to 62 was expected to play their part in the nation at war. Eventually industries were separated into essential for the war effort and non-essential and government developed systems for labour movement between them to increase war production and to feed the population. From December 1916 every avenue was followed to increase the amount of food grown at home by nationalising farming and providing substitutes into agriculture. Farmers were kept on side with minimum crop prices to compensate for their change of land use. Despite the competing demands for manpower, Britain maintained and increased the amounts of home-grown food by prioritising crops to feed the nation, in which the police played an early role as first responders, released as competent substitutes into agriculture, while observing farming and the numerous different sources of substitutes at close quarters.

In order to try to minimise the need for large increases in agricultural manpower, the next chapter considers the government's attempt to expand farm mechanisation. We see the police role with agricultural horses and equipment to increase crop production in the urgency to feed the nation in the second half of the war.

References

1 Journal of the Board of Agriculture, April 1915–March 1916 https://archive.org/details/dli.ernet.232728/mode/2up (accessed 25/1/2024).

2 Journal of the Board of Agriculture, April 1917–March 1918 https://archive.org/details/dli.ernet.29517/mode/2up (accessed 25/1/2024).

3 Barnett, C. 1970. *Britain and Her Army, 1509–1970: A Military, Political and Social Survey*. London: Allen Lane.

4 The Army Council. 1921. *General Annual Reports on the British Army for the Period from 1st October, 1913, to 30th September, 1919*. Cmd. 1193 London: HMSO. Part III Table 1B p. 60.

5 Macready, N. 1925. *Annuls of an Active Life*. vol. I. New York: George H. Doran, p. 245.

6 Armstrong, A. 1988. *Farmworkers: A Social and Economic History 1770–1980*. London: B.T. Batsford.

7 Horn, P. 1984. *Rural life in England in the First World War*. Dublin: Gill and Macmillan.

8 Dewey, P. E. 2014. *British Agriculture in the First World War*. London: Routledge. 2nd edition, pp. 36–37.

9 Ernle, Lord. n.d. *The Land and Its People; Chapters in Rural Life and History*. London: Hutchinson, p. 200.

10 Ministry of Agriculture, A century of Agricultural Statistics 1866–1966 in Whetham E. H. 1978. *The Agrarian History of England and Wales*. Volume VIII 1914–1939. Cambridge University Press p. 44.

11 Dewey, 2014, op. cit. ref 28 p. 57.

12 Dewey, 2014, op. cit. p. 38.

13 Dewey, P. 2000. Chapter 12 Farm Labour. In *The Agrarian History of England and Wales*, ed. E. J. T. Collins. VII (Part 1) 1850–1914. Cambridge: Cambridge University Press, p. 847.
14 Whetham, 1978, op. cit. p. 71.
15 Dewey, 2014, op. cit. pp. 38–9.
16 Montgomery, J. K. 1922. *The Maintenance of the Agricultural Labour Supply in England and Wales During the War*. Rome: The International Institute of Agriculture, ref 4 p. 57.
17 Journal of the Board of Agriculture, August 1915.
18 Crowe, H. 2008. Keeping the Wheels of the Farm in Motion: Labour Shortages in the Uplands During the Great War. *Rural History* 19(2), 201–216, p. 205.
19 The Army Council, 1921, op. cit. Section I p. 6.
20 Wolfe, H. 1923. *Labour Supply and Regulation*. Oxford: Clarendon Press, pp. 20–21.
21 Crowe, 2008, op. cit.
22 Verdon, N. 2017. *Working the Land; a History of the Farmworker in England from 1850 to the Present Day*. London: Palgrave Macmillan.
23 White, B. 2010. Feeding the War Effort: Agricultural Experiences in First World War Devon. *Agricultural History Review* 58, 95–112.
24 Ernle, n.d., op. cit. p. 103.
25 Ernle, n.d., op. cit. p. 116.
26 Whetham, 1978, op. cit. p. 79.
27 Fraser, M. 2019. *Policing the Home Front 1914–1918: The Control of the British Population at War*. Abingdon: Routledge, p. 140.
28 The Army Council, 1921, op. cit. p. 7.
29 McDermott, J. 2011. *British Military Tribunals, 1916–1918: A Very Much Abused Body of Men*. Manchester University Press, https://academic.oup.com/manchester-scholarship-online/book/13569 (accessed 25/1/2024) p. 98.
30 Dewey, 2014, op. cit. p. 58 ref 29.
31 Dewey, P. E. 1984. Military Recruiting and the British Labour Force During the First World War. *The Historical Journal* 27(1), 199–223, p. 215 and 2014, p. 39.
32 MAF Circular A187L: 26 April 1916 (effective from 1 May 1916).
33 Journal of the Board of Agriculture, Circular February 1916.
34 Macready, 1925 op. cit.
35 Tallents, S. 1934. *Man and Boy*. Faber, p. 226.
36 Dewey, P. E. 1984. op. cit. pp. 201–2.
37 TNA MAF 38/180.
38 see HL Parl. Debates, 5th series xxiv, 325–6, 28 Feb 1917 speech by Lord Derby.

39 Dewey, 2014, op. cit. p. 41; Dewey, 1984, op. cit. p. 202.
40 Dewey, 2000, op. cit. p. 823.
41 Journal of the Board of Agriculture, April 1915 pp. 87–90.
42 Department of Agriculture and Technical Instruction for Ireland (14 October 1916). Report and Tables relating to Irish Agricultural Labourers. Agricultural Statistics, Ireland. 1915. Cd. 8386.
43 Dewey, 2014, op. cit. p. 45.
44 Armstrong, 1988, op. cit. p. 161.
45 Young, D. R. 2001. Voluntary Recruitment in Scotland, 1914–1916. PhD University of Glasgow.
46 Board of Education. *School Attendance and Employment in Agriculture. Summary of Returns Supplied by Local Education Authorities for the Period 1st September 1914–31st January 1915*; 31st January 1916; 31st May 1916; 16th October 1916.
47 Horn, 1983, op. cit. p. 207, Table 1.
48 Dewey, P. E. 1979b. Government Provision of Farm Labour in England and Wales, 1914–1918. *The Agricultural History Review* XXVIII, 110–21, pp. 112, 117–8
49 Horn, 1983, op. cit. p. 210.
50 Brown, N. F. 2015. 'Fall In the Children': A Regional Study of the Mobilisation of the Children of the 42nd Regimental Area During the Great War. PhD thesis, University of Dundee, Department of History,
51 Whetham, 1978, op. cit. p. 72.
52 HC Deb 25 February 1915 vol 70 cc402–05.
53 Horn, 1983, op. cit. p. 206.
54 Green, F. E. 1920. *A History of the English Agricultural Labourer 1870–1920*. London: P. S. King, pp. 235–239.
55 Horn, 1983, op. cit. p. 212.
56 Ernle, n.d. op. cit. p. 128.
57 Dewey, 2014, op. cit. p. 51 and 129; ref 111 and 112 p. 145.
58 Dewey, 2014, op. cit. Appendix G.
59 Ernle, n.d. op. cit. p.177–9.
60 Middleton, T. H. 1923. *Food Production in War.* Oxford: The Clarendon Press.1923, op. cit. p. 223.
61 Dewey, 2014, op. cit. ref 53 p. 59.
62 Whetham, 1978, op. cit. p. 80.
63 Dewey, 2014, op. cit. p. 59 refs 57 and 59.
64 Dewey, 2014, op. cit. pp. 52–5.
65 Dewey, 2014, op. cit. p. 56 and ref 72 p. 59.
66 Whetham, 1978, op. cit. p. 81.
67 Dewey, 2014, op. cit. p. 48.
68 Dewey, 2014, op. cit. pp. 49–50.

69 Dewey, 2014, op. cit. pp. 49–50 and ref 43 p. 58.
70 McDermott, 2011, op. cit. pp. 106–7.
71 Whetham, 1978, op. cit. p. 82.
72 Dewey, 2014, op. cit. p. 113.
73 Dewey, 2014, op. cit. p. 56.
74 Montgomery, 1922, op. cit.
75 Dewey, 2014, op. cit. pp. 46–7.
76 Dewey, 2014, op. cit. p. 85.
77 see Dewey, 1984, op. cit. Table 2, p. 204.
78 Dewey, 1984, op. cit. p. 218.
79 Dewey, 2014, op. cit. p. 39.
80 Montgomery, 1922, op. cit.
81 Dewey, 2014, op. cit. Table 4, p. 208.
82 Dewey, 2014, op. cit. p. 44.
83 Montgomery, 1922, op. cit. pp. 9–10.
84 McDermott, 2011, op. cit. p. 117.
85 Middleton, 1923, op. cit. p. 140.
86 Journal of the Board of Agriculture, March 1916 p. 1301.
87 Montgomery, 1922, op. cit. p. 8.
88 Journal of the Board of Agriculture, 1917 April p. 99.
89 Dewey, 2014, op. cit. pp. 140–1.
90 The War Cabinet. 1918. Report for the year 1917. Cd. 9005 HMSO. https://archive.org/details/reportforyear19100grea/mode/2up?view=theater (accessed 25/1/2024), pp. 83–4.
91 Troup, Agricultural census, 1916, TNA HO 45/10827 letter dated 7th November 1916.
92 Montgomery, 1922 op. cit. p. 12.
93 Letter 322,579/2 from the Home Office, 7th November 1916. TNA HO 45/10827/322579.
94 Fraser, 2019, op. cit. p. 196.
95 *The Newcastle Daily Journal* Feeding Stuffs and Livestock. 14 April 1917 p. 12.
96 *The Newcastle Daily Journal* 19 April 1917 p. 6.
97 Norfolk War Agricultural Committee 2 November 1917 C/C 10/16 Norfolk Record Office, Norwich.
98 Journal of the Board of Agriculture, June 1917 p. 341
99 Journal of the Board of Agriculture, October 1916, pp. 699–701; December 1916, pp. 900, 903–904.
100 McDermott, 2011, op. cit. p. 111.
101 Montgomery, 1922, op. cit. p. 12.
102 *The Birmingham Daily Mail* 13 April p. 3.
103 Montgomery, 1922, op. cit. p. 10.

104 Middleton, 1923, op. cit. p. 180.
105 Montgomery, 1922, op. cit. p. 11.
106 MAF Circular A247C (5 October 1916); LGB Circular R.102 (6 October 1916), para. 5 in McDermott, 2011, p. 110.
107 Grieves, K. 1988. *The Politics of Manpower, 1914–18*, 91. Manchester: Manchester University Press.
108 Grieves, 1988, op. cit. pp. 74 and 86.
109 TNA CAB 23/1 W.C. 19 January 1917.
110 Grieves, 1988, op. cit. pp. 99–100.
111 Journal of the Board of Agriculture, 1917, p. 1173.
112 Montgomery, 1922, op. cit. pp. 24–25.
113 Lloyd George MSS, F/15/8/3, Prothero to Lloyd George, 11 January 1917.
114 TNA CAB 23/1 W.C. 39 19 January 1917; W.C. 40 22 January 1917.
115 Barnett, 1985, op. cit. p. 200.
116 TNA CAB 23/1 W.C. 55.
117 Grieves, 1988, op. cit. pp. 106–7.
118 TNA CAB 23/1 W.C.45 25 January 1917.
119 Grieves, 1988, op. cit. p. 99.
120 Dewey, 2014, op. cit. pp. 113–4; refs 28–30.
121 Ernle, n.d., op. cit. p. 125.
122 Middleton, 1923, op. cit. p. 182.
123 Middleton, 1923, pp. 183–4.
124 *The Police Review and Parade Gossip*, 18 February 1916 "The Military Service Act, 1916".
125 TNA MH 47/142/2, List of Certified Occupations, 1st February 1917 R117. p. 20.
126 Journal of the Board of Agriculture, June 1917 pp. 339–341.
127 Dewey, P. E. 1979a. Food Production and Policy in the United Kingdom, 1914–1918. *Transactions of the Royal Historical Society* 30, 71–89, p. 78.
128 Middleton, 1923, op. cit. p. 185.
129 Dewey, 2014, op. cit. p. 118.
130 Dewey, 2014, op. cit. p. 118 ref 56 p. 143.
131 *The Scotsman* Military Men. Difficulties with Men. 16 March 1917 p. 6.
132 Dewey, 2014, op. cit. p. 114 ref 31 p. 143.
133 Agricultural Wages Board. 1919. *Report on Wages and Conditions of Employment in Agriculture*. Board of Agriculture and Fisheries. Cmd. 24 Vol IX, p. 49.
134 Armstrong, 1988, op. cit. p. 162.
135 Whetham, 1978, op. cit. p. 100.
136 Dewey, 2014, op. cit. p. 114 ref 34 p. 143.
137 Middleton, 1923, op. cit. p. 185.

138 Ernle, n.d., op. cit. p. 143.
139 TNA CAB 23/3/17 pp. 6–7.
140 Dewey, 2014, op. cit. p. 187.
141 Agricultural Wages Board, 1919, op. cit. pp. 121–3.
142 Dewey, 2014, op. cit. p. 188.
143 Board of Agriculture for Scotland. 1918. *Sixth Report*, for the Year Ending 31st December 1917. Cd. 9069. Edinburgh HMSO. National Archives of Scotland AF 78/6, p. li.
144 Dewey, 1979b, op. cit. p. 112.
145 Dewey, 1979b, op. cit. p. 112.
146 LGB Circular R.122, 1 March 1917, in McDermott, 2011, p. 26.
147 *House of Commons Debates HC05 March 1917 vol 91 cc65–6.*
148 Ernle, n.d., op. cit. p. 200.
149 Crowe, op. cit. p. 207.
150 Journal of the Board of Agriculture, April 1917 pp. 120–1.
151 Dewey, 2014, op. cit. p. 114 ref 38 p. 143.
152 Ernle, n.d., op. cit. p. 127.
153 Middleton, 1923, op. cit. p. 186.
154 Dewey, 2014, op. cit. pp. 114–5 and ref 39.
155 Dewey, 2014, op. cit. pp. 116–7 and ref 44 and 49–51 p. 143.
156 Dewey, 1979b, op. cit. p. 112.
157 Middleton, 1923, op. cit. p. 231.
158 Dewey, 2014, op. cit. p. 117 ref 49–51 p. 143.
159 Dewey, 2014, op. cit. pp. 130–3.
160 Ernle, n.d., op. cit. p. 178.
161 Ernle, n.d., op. cit. p. 181.
162 Ernle, n.d., op. cit. pp. 172–3.
163 Middleton, 1923, op. cit. p. 186.
164 Ernle, n.d., op. cit. pp. 128 and 173.
165 Dewey, 2014, op. cit. pp. 130–2.
166 Ernle, n.d., op. cit. pp. 170–1.
167 Dewey, 2014, op. cit. p. 132 and ref 120 p. 154.
168 Journal of the Board of Agriculture, March 1917, pp. 1282–3.
169 Dewey, 2014, op. cit. p. 131 and refs under Table 9.8.
170 Ernle, n.d., op. cit. pp. 128–9.
171 Montgomery, 1922, op. cit. p. 13.
172 Dewey, 2014, op. cit. pp. 134–5 ref 132, 134, 136 and 138 p. 14.
173 Charman, T. 2015. *The First World War on the Home Front*. London: Andre Deutsch, p. 158.
174 Dewey, 1979b, op. cit. p. 116.
175 Ernle, n.d., op. cit. p. 127.

176 Board of Agriculture and Fisheries. *Wages and Conditions of Employment in Agriculture I General Report* PP 1919 Cmd. 24, vol IX pp. 58–9.
177 *The Scotsman* 3 March 1917 p. 7.
178 Dewey, 2014, op. cit. p. 48.
179 Agricultural Wages Board. 1919. *Report on Wages and Conditions of Employment in Agriculture*. Board of Agriculture and Fisheries. Cmd. 24 Vol IX, para 149.
180 Ernle, n.d., op. cit. p. 129.
181 Middleton, 1923, op. cit. p. 186.
182 Horn op. cit. pp. 212–4.
183 Brown, 2015 op. cit. pp. 178–80.
184 Dewey, 2014, op. cit. p. 120.
185 Ernle, n.d., op. cit. p. 127.
186 Dewey, 2014, op. cit. pp. 120–22.
187 Crowe op. cit. p. 213.
188 TNA CAB 23/4/71.
189 Dewey, 2014, op. cit. pp. 122–3 ref 82–3 p, 144.
190 Ernle, n.d., op. cit. p. 145.
191 Dewey, 2014, p. 123 and ref 84 p. 144.
192 Dewey, 2014, op. cit. pp. 122, 144, ref 81, and 191.
193 Dewey, 2014, op. cit. p. 125–7 and ref 91 p. 144.
194 Dewey, 2014, op. cit. p. 122.
195 TNA NATS 1/132.
196 Dewey, 1979b, op. cit. pp.112–3.
197 The War Cabinet, 1918, op. cit. p. 81.
198 Dewey, 2014, op. cit. p. 117 and ref 48 and p. 118 ref 54 p. 143.
199 Ernle, n.d., op. cit. p. 143.
200 Ernle, n.d., op. cit. p. 130.
201 Ernle, n.d., op. cit. p. 107.
202 Barnett, 1985, op. cit. p. 194.
203 Middleton, 1923, op. cit. p. xii.
204 Grieves, 1988, op. cit.
205 The War Cabinet, 1918, op. cit.
206 Ernle, n.d., op. cit. p. 106–7.
207 *The Daily News* 21 December 1916 pp. 1–2.
208 Middleton, 1923, op. cit. p. 165.
209 Journal of the Board of Agriculture, January 1917, pp. 1008–9.
210 Journal of the Board of Agriculture, January 1917, pp. 1009–11.
211 Journal of the Board of Agriculture, January 1917, pp. 1012–3
212 Middleton, 1923, op. cit. pp. 165–6.
213 The Royal Society. 1916. *The Food Supply of the United Kingdom* PP 1916 Cd 8421.

214 Whetham, 1978, op. cit. p. 93.
215 Dewey, 2014, op. cit. p. 96.
216 Gregory, A. 2008. *The Last Great War: British Society and the First World War*. Cambridge: Cambridge University Press.
217 Journal of the Board of Agriculture, 1917, pp. 1121–26.
218 Dewey, 2014, op. cit. pp. 97 and 171.
219 Ernle, n.d., op. cit. pp. 108–9.
220 Middleton, 1923, op. cit. pp. 167.
221 Dewey, 2014, op. cit. p. 171.
222 *Chester Chronicle* 3 March p. 3.
223 Scottish Departmental Committee on Food Production, Second Report. 1917. *Production of Food in Scotland*. London: HMSO, p. 5. Scottish Archives HH31/24/6.
224 Scottish Departmental Committee on Food Production, Third Report. 1917. *Production of Food in Scotland* London: HMSO, p. 3. Scottish Archives HH31/24/12.
225 Conacher, H. M. 1926. Increased Food Production. Chapter IX. In *Rural Scotland During the War*, ed. Jones et al. London: Humphrey Milford, Oxford University Press, pp. 122–86, p. 166.
226 Scottish Departmental Committee on Food Production, Third Report (1917) op. cit. p. 4.
227 Dewey, 2014, op. cit. p. 96.
228 Middleton, 1923, op. cit. pp. 172–5.
229 Ernle, n.d., op. cit. p. 111.
230 Dewey, 2014, op. cit. pp. 96–7.
231 Ernle, n.d., op. cit. p. 118.
232 Middleton, 1923, op. cit. p. 169 and 221.
233 Farmers, The Army, and Our Food Supply. *The Weekly Bulletin* March 1, 1917, p. 26.
234 Ernle, n.d., op. cit. pp. 138–9.
235 Rew, 1918, op. cit. p. 44.
236 Whetham, 1978, op. cit. p. 91.
237 Dewey, 2014, op. cit. p. 92.
238 Dewey, 2014, op. cit. p. 93 see ref 9 p. 104.
239 Dewey, 2014, op. cit. pp. 171–2.
240 Dewey, 2014, op. cit. p. 98 and 172–4.
241 Dewey, 2014, op. cit. p. 179 and ref 23 p. 196.
242 Ernle, n.d., op. cit. p. 110.
243 Prothero, R. E. 1938. *Whippingham to Westminster*. London: John Murray, p. 307.
244 Barnett, 1985, op. cit. p. 198.
245 Whetham, 1978, op. cit. p. 98.

246 Whetham, 1978, op. cit. pp. 92–4.
247 Dewey, 2014, op. cit. pp. 95–6 and 179 ref 24 p. 196.
248 Middleton, 1923, op. cit. p. 171.
249 Middleton, 1923, op. cit. p. 289.
250 Dewey, 2014, op. cit. p. 180/1.
251 Dewey, 2014, op. cit. p. 175.
252 Middleton, 1923, op. cit. p. 218.
253 Barnett, 1985, op. cit. p. 198.
254 Journal of the Board of Agriculture, April 1917 pp. 111–2.
255 Dewey, 2014, op. cit. p. 112.
256 *Birmingham Gazette* 13 April 1917 p. 2.
257 Dewey, 2014, op. cit. p. 111 and ref 21 p. 142.
258 Middleton, 1923, op. cit. p. 172.
259 Ernle, n.d., op. cit. p. 110.
260 *The Diss Express, and Norfolk and Suffolk Journal* 12 January 1917 p. 8.
261 Journal of the Board of Agriculture, June 1917 pp. 315–21.
262 Dewey, 2014, op. cit. p. 99 and p. 181/2 and ref 31 p. 196.
263 Middleton, 1923, op. cit. pp. 212–4.
264 TNA WO 162/28 Supply of Men for the Army, Memorandum by the Military Members of the Army Council to the Secretary of State for War (Lloyd George), 28 November 1916.
265 Grieves, 1988, op. cit. pp. 64–5.
266 Grieves, 1988, op. cit. pp. 68–9.
267 Grieves, 1988, op. cit. pp. 70–1.
268 *HC Deb 19 December 1916 vol 88 cc1353.*
269 Grieves, 1988, op. cit. p. 73.
270 Grieves, 1988, pp. 75–6, The War Cabinet. Report for the year 1917. Cd. 9005 p. 8.2.
271 *HC Deb 19 December 1916 vol 88 cc1349–51.*
272 Grieves, 1988, op. cit. p. 76–9.
273 Macready, 1925, op. cit. p. 255.
274 Grieves, 1988, op. cit. p. 94.
275 Parliamentary Archives GB-061 LG/F/15/8/3. 11 January 1917.
276 Grieves, 1988, op. cit. pp. 109–10.
277 *Manchester Guardian* 7 February 1917 p. 5.
278 *The Times* 7 February 1917 p. 9.
279 Middleton, 1923, op. cit. p. 186.
280 Grieves, 1988, op. cit. p. 112.
281 Dewey, 2014, op. cit. p. 138 ref 155.
282 Ernle, n.d., op. cit. pp. 126–7.

283 See for example: *The Times*, 28 March 1917 p. 3. *The Glasgow Herald*, 28 March 1917 p. 6; *The Evening Times*, 28 March 1917 p. 3; *The Evening News*, 28 March 1917 p. 4.
284 *HC Deb 02 April 1917 vol 92 c887.*
285 *The Scotsman* 26 March 1917 p. 6.
286 National Service Report No. 20 for Week Ending 7th June 1917, author Neville Chamberlain. TNA CAB 24/16/7.
287 Macready, 1925, op. cit. p. 264.
288 *HC Deb 14 February 1917 vol 90 c620; HC Deb 15 February 1917 vol 90 cc914–26.*
289 Grieves, 1988, op. cit. pp. 72–3.
290 Winter, J. M. 1985. *The Great War and the British People*. Basingstoke: Palgrave Macmillan, p. 41.
291 Dewey, 1979b, op. cit. p. 119–20.
292 *HC Deb 22 February 1917 vol 90 cc1543–44.*
293 Ernle, n.d., op. cit. p. 69.
294 Dewey, 1979b, op. cit.
295 TNA CAB 23/3/32 184 p. 5.
296 Dewey, 1979b, op. cit. p. 112.
297 *The Weekly Bulletin*. Farmers, the Army, and Our Food Supply 1 March 1917 p. 26.
298 Armstrong, 1988, op. cit. p. 168.
299 Grieves, 1988, op. cit. p. 137.
300 The War Cabinet, 1918, op. cit. pp. 85–6; Wolke, 1923, pp. 50–6.
301 Grieves, 1988, op. cit. p. 148.
302 CAB 24/28, GT 1481, A. C. Geddes, The Theory and Practice of Recruiting, 22nd July 1917.
303 Winter, 1985, op. cit. p. 42.
304 TNA NATS 1/338 Minutes of the National Service Council 12 September 1917; TNA CAB 23/4 W.C. 231 Functions of the Ministry of National Service, 12 September 1917.
305 Grieves, 1988, op. cit. p. 176.
306 TNA CAB 24/28/95.
307 Winter, 1985, op. cit. p. 45.
308 HC Deb 14 January 1918 vol 101 cc86.
309 McDermott, 2011, op. cit. p. 27.
310 Journal of the Board of Agriculture, June 1917 p. 341.
311 Journal of the Board of Agriculture, July 1917 pp. 455–6.
312 Middleton, 1923, op. cit. pp. 214–8.
313 Dewey, 2014, op. cit. p. 99 and ref 33 p. 105.
314 Ernle, n.d., op. cit. p. 142.
315 Journal of the Board of Agriculture, August 1917 pp. 565–6.

316 Dewey, 2014, op. cit. p. 99.
317 Ernle, n.d., op. cit. pp. 126–7.
318 Journal of the Board of Agriculture, July 1917 pp. 465–6.
319 Middleton, 1923, op. cit. pp. 208–18.
320 Whetham, 1978, op. cit. p. 107.
321 *HC Deb 14 January 1918 vol 101 cc58–86.*
322 Middleton, 1923, op. cit. pp. 234–5.
323 Military Service (No.2) Act of March 1918 Ch. 5. June 1918.
324 Dewey, 2014, op. cit. p. 107.
325 Middleton, 1923, op. cit. p. 265.
326 Dewey, 2014, op. cit. p. 109.
327 Dewey, 2014, op. cit. p. 102.
328 TNA CAB 24/52, 422(1), GT 4658, 30 May 1918.
329 Middleton, 1923, op. cit. p. 265.
330 Crowe op. cit. p. 206.
331 HC Deb 11 June 1918 vol 106 cc2019–20; HL Deb 08 August 1918 vol 31 cc673–8.
332 Whetham, 1978, op. cit. p. 112.
333 Middleton, 1923, op. cit. p. 237.
334 Whetham, 1978, op. cit. p. 108.
335 Dewey, 1979b, op. cit. pp. 111–2.
336 Ernle, n.d., op. cit. pp. 161–3 and 5–6.
337 Dewey, 2014, op. cit. p. 103.
338 Whetham, 1978, op. cit. pp. 118–9.
339 McDermott, 2011, op. cit. p. 29.
340 Salter, J. A. 1921. *Allied Shipping Control: An Experiment in International Administration*. Oxford: The Clarendon Press, pp. 122–6.
341 Dewey, 1978, op. cit. p. 80.
342 Middleton, 1923, op. cit. p. 286.
343 Dewey, 1979a, op. cit. p. 81.

of foodstuffs. Food for animals was diverted to feed humans and farmers advised that with the high price and shortage of oats they should feed horses on substituted dried grains, bran, maize and linseed cake instead [21]. From April 1917 compulsory restrictions on horse feed were issued (The Horse Rationing Order) in force from May 1917 [22]. Whereas surplus cattle not useful for milk production or breeding were culled as the use of land for grazing was less efficient than arable to grow crops (see Chapter 3, p. 61)[3] and they competed with humans for grain, it was essential to feed horses to maintain or increase their numbers for agricultural use. However, with decreasing food products, further restrictions by March 1918 due to the shortage of oats, maize and other cereals led the Food Controller to ration horses to a diet with bran and dried brewers' grain with advice on adjustments to the amount of work they were expected to perform [23, 24]. In May 1918 the army also requisitioned all hay for army horses [25].

Police involvement in agriculture allowed surveillance that horses were fed according to rations and farmers were prosecuted for feeding oats to their horses. By June 1918 army horses were also being rationed to a combination of oats combined with maize [26]. A further incidental restriction on the use of horses in agriculture came from the loss of skilled manpower by 1917, which led some horses to be unshod, making them less efficient on heavier soil [27, 28].

To supplement the use of horses, help for farmers by mechanisation in agriculture started around 1830 with the introduction of steam ploughs, but equipment on arable land for ploughing, reaping and threshing, introduced to Britain in the Great Exhibition of 1851, was sparse [29]. Agricultural machinery use began slowly from 1866 when larger farms started to use steam ploughs with reaping and threshing equipment, the

3

100 acres of very poor grass fed to cattle	could feed 2–3 people for 1 year
100 acres of medium grass fed to cattle	could feed 12–14 people for 1 year
100 acres of very good grass fed to cattle	could feed 25–40 people for 1 year
100 acres of fodder beet (mangolds)	could feed 35 people for 1 year
100 acres of wheat as bread	could feed 200 people for 1 year
100 acres of potatoes	could feed 400 people for 1 year

latter powered by steam or a horse wheel, particularly in the Scottish Lowlands, Northumberland and Lancashire [30], although the number of sets are not known [31]. However, steam ploughs were used less for ploughing in Scotland on light soil where improved horse-drawn ploughs were preferred [32], although steam ploughs with threshing machines, even on the smallest farms since the turn of the century, decreased the need for labour [33]. By 1917 steam ploughing was sufficiently developed and in use on larger farms, that the Agricultural Machinery Board of the Ministry of Munitions submitted an order to John Fowler and Company of Leeds for 140 steam engines with balance ploughs, cultivators, water carts and living vans for contractors working the equipment [34]. Due to the high cost of steam engines which could not be used for other farm work, most steam ploughing was in the hands of contractors who travelled from farm to farm [35]. Steam ploughs were mainly used for breaking up previously uncultivated soil to destroy deep-rooted shrubs and bushes that manual ploughing with horses could not handle [36–38].

The use of steam ploughs also suffered from loss of skilled labour by 1917, despite drivers, attendants and mechanics receiving exemption from military service in 1916 (see Chapter 3, p. 47) [39]. Reports showed that only around half of the 500 steam plough sets in use in Britain before 1916 were still in operation due to skilled manpower shortages. The FPD ordered "65 new sets, 6,000 ploughs, 5,500 sets of harrows, 5,000 binders and 3,365 rollers" [40], but this was a very small number compared with the total amount of farm machinery. The army supplied 300 experienced men to repair existing sets which were fit for use by June 1917 and the Steam Cultivation Association also agreed to keep the sets working 7 days a week during daylight hours, resulting in 1,200,000 acres ploughed by steam ploughs in preparation for the 1918 harvest [41, 42]. But despite these efforts, by December 1917 around half the steam plough sets were idle for lack of repair, skilled men to operate them or coal [43, 44].

With the invention of the combustion engine several farm implements were developed and introduced into farming from 1900 [45]. Attempting to reduce the need for manpower, the Board of Agriculture boosted mechanisation by promoting the use of petrol-driven tractors in November 1915 [46], but the conventional nature of farmers made them sceptical of their utility, due to their novelty [47]. Furthermore, the Milner Report of 1915 revealed insufficient labour and materials hindered the production of tractors and ploughs in Britain [48]. Although receiving farmers' attention from January 1916, the capabilities of petrol-driven tractors were unknown and the numbers in use in the UK by the end of 1916 were only around 3,500 [49]. Furthermore, shortages of petrol began, increasing the role of the WAECs in rationing and distribution.

The stimulus to increased mechanisation occurred from around 18th October 1916 when the sinking of ships and inevitable loss of food imports led to the need for efficient and expanded home food production to pacify the population and avert imminent starvation (see Chapter 2). The change of government in December 1916 heralded wide-ranging interventionist policies (see Chapter 3), to increase arable production when much tilling was manual using hand ploughs and horses. Most farmers were willing to comply with directives from the WAECs, but were unable to do so without labour, horses and feeding stuffs, machinery and fertilisers [50, 51]. In Scotland, the Wason Committee also considered how to increase arable production in their second report of 1917 affirming farmers' willingness, but saying no substitute manpower was available, although policemen had been released to help from March in many areas (see Chapters 7 and 8). Their second report recommended increased mechanisation by motor and horse-drawn ploughs [52].

WAECs required sub-committees and District Committees to ensure farmers had access to sufficient labour, horses, mechanical supplies such as petrol-driven tractors and spare parts, coal for cultivation with steam ploughs and other essential supplies [53, 54] (see also Chapter 3, p. 62). WAECs reported their needs to the Food Production Department (FPD). On 23rd December 1916 in response to increasing demand the Treasury agreed to purchase tractors from America [55], although the FPD through the WAECs warned farmers that the "agricultural tractor at its present stage of development is the least reliable of all ploughing instruments." In early 1917 WAECs took charge of privately owned petrol-driven tractors arranging to share them between farms, supplemented by publicly owed tractors in April, and in May by machines bought by the FPD mainly from America [56].

From January 1917 the order from America, estimated to reach 4,000 in 1917 and 7,000 by 1919, started to arrive [57] but in the first few months of 1917 the delivery of tractors and machinery including horse-drawn ploughs was slow and hindered local implementation of the plough policy [58]; many farmers complained in early 1917 that they had not been supplied with the equipment to plough the additional land ordered by the WAEC [59]. Canada also promised to donate 100 tractors with harrows and binders, including a skilled mechanic with each [60]. By the end of May 1917, 666 were at work in England and Wales, averaging around 12–13 per county, lent to farmers on request, but distribution was uneven [61, 62]. By the end of 1917 the FPD estimated around 3,500 privately owned petrol-driven tractors were at work in the UK and 1,550 bought by government mainly from America [63], WAECs shared them between farms [64]. But the supply of petrol-driven tractors and spare parts fell well short of need and orders, which limited the 1918 ploughing targets; they failed to arrive in large numbers from America until February 1918 [65, 66]. By the end of November 1917 only 1,500 had been supplied and by March a further 1,000 from the order for 5,000 on which ploughing targets had been based [67]. By June 1918 3,000 Fordson tractors had arrived from America and were ready for use, with a further 3,000 to be delivered [68]. They became the most popular model.

The slow delivery of tractors complicated spring ploughing in 1918 in most counties [69–71]. Their experimental nature on British farms and the many different types and models by mid-1918 also gave rise to problems of the supply of spare parts [72]. Their untested nature meant that many were found to be unsuitable for different types of soil and continually broke down [73]. An example from Worcestershire showed that for one year from mid-September 1917, tractors were idle for more than 50% of the time, also some areas, such as Bedfordshire, claimed they were unsuitable for most soil [74]. These problems with delivery and maintenance emphasised the need for manual ploughing with horses.

In early 1917 publicly owned tractors were still under the control of the FPD, rather than devolved to each WAEC, only in mid-1918 were they decentralised to WAECs, who developed sub-committees with specialist mechanical expertise. In August 1917 the FPD delegated distribution of tractors and implements to a Machinery Sub-Committee of each WAEC with a local machinery officer who decided which land was suitable for tractor ploughing and agreed contracts with farmers. As predicted by the FPD, tractors frequently broke down [75] mainly due to their allocation to the heaviest work. In addition, around 30 different models were supplied, many of which were novel to the mechanics. The police lent to farms also had sparse knowledge of petrol-driven machines, limited to their use of motorcycles, particularly by dispatch riders. Therefore, use and maintenance of petrol-driven machinery was also a novelty in the police. Police cars were rare, the first bought to transport the

Chief Constable and senior officers were introduced slowly around Britain before 1914. Training to drive and maintain cars was novel and patchy across Britain before the end of the war[4] therefore police help with petrol and paraffin-driven agricultural machines would have been very limited.

However, the police were involved in traffic regulation from 1902 bringing charges against contravention of local regulations [76, 77] and traffic accidents [78]. The police were also aware of the further restricted use of early tractor models, as the iron strakes on the solid wheels destroyed road surfaces confining them to pastureland. The combined disadvantage with the use of these machines led to only half the area ploughed by steam ploughs being ploughed by petrol-driven tractors in the 1917–18 ploughing campaign.

In order to build capacity, the lack of skilled manpower for horse and mechanical ploughing was addressed by the establishment of ploughing schools from September 1917 which gave 2 weeks training, some farm training centres were also established. Combined, these centres trained around 4,000 ploughmen in manual ploughing with horses [79], 4,000 in tractor ploughing and 200 in steam ploughing [80]. Furthermore, some farmers were willing to train substitute soldiers in ploughing and general farm work when they demonstrated a willingness to learn, particularly those who were deployed permanently until the end of the war [81]; by November many WAECs had established training schools for horse ploughmen, and others had sent men to be trained by farmers [82]. The number of men able to operate petrol-driven tractors was also increased when training was given for 2 weeks from early 1917. Later WAECs sent one man for each vehicle to be trained by the local tractor representative. In total 4,093 men were trained in petrol-driven tractor ploughing [83], also by May 1918 women started to receive training to use Fordson tractors [84]. However, of the mechanical means of cultivation, steam ploughs were more successful at breaking up heavy and water-logged soil and removing thorn and other stubborn bushes from derelict land [85, 86].

[4] I am most grateful for private communication on the history of police motorcycles and cars and their introduction into different police forces in Britain from members of the Police History Society: Roger Young, Pam Mills, Sabine Skae, Fred Feather, Keith Jackson, Raymond Lee, Robert Bartlett, John Dowell, Terry Gardner, Steve Bridge, Alison Jay, Alan Moss, Mark Rothwell and Rebecca Walker. Also, for the private communication from Paddy Carpenter, archivist of the Police Vehicle Club.

The novelty of tractor ploughing gave the additional benefit of raising public morale by publicising their use and involved the police in regulation. Emphasis on the urgency for ploughing and planting crops in early 1917 led to 24-hour tractor ploughing 7-days a week lit by acetylene lamps at night in some areas, for example Cheshire [87] and the north of England on land adjacent to a munition factory [88]:

> "NIGHT PLOUGHING". The urgency of ploughing and planting spring crops was seen in scheduling the motor ploughs to be worked in 3 8-hour shifts: 12midnight – 8am; 8am-4pm; 4pm-12midnight, 7 days a week. If tractors were used, they could plough for 7 hours in each shift allowing time for maintenance. Estimates were that 3 acres (approximately the size of 3 football pitches) could be ploughed per 7-hour shift; the working period would be extended to 40 days and nights which would make it possible to plough 360 acres per tractor. 1 of the tractors arrived on Wednesday 7th March. During night ploughing acetylene headlights were used "the police offering no objections to these lights, so long as they are properly shaded when travelling along the roadway." [89]

The above quotation emphasises the urgency for ploughing and how the novel petrol-driven tractors were being deployed to maximum capacity. It shows readers the anticipated increase in acreage of ploughing and planting crops over traditional manual ploughing with horses in reports aimed to raise public morale. The involvement of the police in mechanical ploughing was to ensure lighting did not infringe the requirements for shading from enemy aircraft at night. The novelty value of petrol-driven tractors and their public image was seen in similar reports nationwide, watching the tractors in action became a public spectacle and confidence booster. The Journal of the Board of Agriculture forecast and promoted night ploughing as early as 1915 to make the most use of fine weather and as labour-saving [90].

An early petrol-driven tractor with ploughing equipment is shown in Fig. 4.1.

Rowland Prothero, President of the Board of Agriculture from December 1916, attempted to increase public involvement in agriculture throughout Britain: anticipating religious objection to Sunday ploughing, he gained agreement from the Archbishop of Canterbury to remove religious protest: "I have no hesitation in saying that in the need which these weeks present men and women may with a clear conscience do field work on a Sunday" [91]. A further suggestion to appease regular churchgoers

Fig. 4.1 From *The Sheffield Daily Independent*, 3 March 1917 p. 5

was for the clergy to restrict services to only one on a Sunday evening, so that the maximum number of people could continue with ploughing and planting during daylight hours.

By October 1917 some of the uses and disadvantages of different types of tractors were becoming evident and farmers were warned of their appropriate use in different agricultural conditions, such as on boggy or stony ground, where horses were preferable [92]. Even by November 1918 when many had been in use for more than a year, petrol-driven tractor repairs occupied around 16% of the time in England and Wales on nearly 3,000 government-provided machines [93]. These problems resulted in steam tackle and horse-drawn ploughs being the most useful and reliable ploughing methods [94]. However, despite these problems, from October 1917 petrol-driven tractor ploughing accounted for some 480,000 acres of ploughed land, increasing to 600,000 acres during spring 1918 in preparation for sowing [95, 96]. But their numbers by the end of the war did not amount to more than around 25,000—on aggregate only one for around 16 farms, shared between farms [97].

In conclusion, wartime conditions and the need for urgent home crop production to alleviate imminent population starvation gave the police increased involvement with horse and mechanical means of agricultural production. The number of horses increased to pre-war numbers by the time of the plough policy, but the skilled manpower able to plough was very limited. Police experience with horses combined with previous agricultural work would have provided welcome help to farmers trying to comply with government ploughing targets. The police would also have advised and cautioned farmers on restrictions to horse feed, when these were imposed mid-1917, by which time many policemen were allocated to farms. Access to farmers during distribution and collection of the government survey of horses in April 1917 provided further opportunities to advise and warn farmers of their responsibilities towards horses and later the restrictions on their sale.

Labour shortages in agriculture also led farmers to experiment with machinery under the control of the Food Production Department, despite initial scepticism. The role of policemen with mechanical farming equipment was to ensure adherence to regulations, particularly in road transport and shading lights from enemy aircraft at night. In this work they supported targets set by the Ministry of Agriculture and the Food Production Department implemented locally.

Although much effort was made to increase mechanical help for farmers to minimise the need for manpower, much petrol-driven equipment was underdeveloped and made little impact, although tested for endurance on different soil types. Delivery from the manufacturers was also late, which hindered ploughing targets and frustrated farmers.

But how did the food production programme affect the large majority of the population and help to allay threats of starvation? The next chapter shows the traditional diets of the masses and the effects on them of food privation during the war, particularly in the years 1917–1918. It also shows the growing unrest as food shortages and huge price inflation grew which threatened to destabilise the production of war materials at home. We also see government efforts to conserve staple food items such as bread and sugar before rationing was considered necessary.

References

1. Journal of the Board of Agriculture, April 1915–March 1916 https://archive.org/details/dli.ernet.232728/mode/2up (accessed 25/1/2024).
2. Journal of the Board of Agriculture, April 1917–March 1918 https://archive.org/details/dli.ernet.29517/mode/2up (accessed 25/1/2024).

3 Lloyd George, D. 1934. *War Memoirs of David Lloyd George* III. London: Ivor Nicholson, p. 1302.
4 Ministry of Agriculture, Fisheries and Food, Department of Agriculture. 1986. *A Century of Agricultural Statistics: Great Britain 1866–1966*. London: HMSO, p. 70.
5 Ministry of Agriculture, Fisheries and Food. Department of Agriculture and Fisheries for Scotland. 1968. op. cit. pp. 54–5.
6 HC Deb 06 August 1914 vol 65 cc2064–5.
7 HC Deb 08 August 1914 vol 65 cc2203–4; HC Deb 10 August 1914 vol 65 cc2262–3.
8 Whetham, E. H. 1978. *The Agrarian History of England and Wales* Volume VIII 1914–1939. Cambridge: Cambridge University Press, p. 73.
9 Armstrong, A. 1988. *Farmworkers: A Social and Economic History 1770–1980*. London: B.T. Batsford, p. 156.
10 Journal of the Board of Agriculture, June 1915 p. 276; March 1916 p. 1254.
11 Ministry of Agriculture, Fisheries and Food, Department of Agriculture. 1986. *A Century of Agricultural Statistics: Great Britain 1866–1966*, op. cit., Table 70 p. 129.
12 Journal of the Board of Agriculture, July 1916 p. 400.
13 Journal of the Board of Agriculture, May 1917 pp. 234–5.
14 Journal of the Board of Agriculture, August 1917 pp. 575–9.
15 Middleton, T. H. 1923. *Food Production in War*. Oxford: The Clarendon Press, p. 227.
16 Dewey, P. E. 2014. *British Agriculture in the First World War*. London: Routledge. 2nd edition, p. 115 and ref 40 p. 143.
17 Barnett, M. L. 1985. *British Food Policy During the First World War*. Boston: George Allen and Unwin, p. 200.
18 TNA HO 45/10840/333647.
19 HC Deb 12 July 1917 vol 95 c2116.
20 Journal of the Board of Agriculture, July 1917 pp. 471–2.
21 Journal of the Board of Agriculture, March 1917 p. 1261.
22 Dearle, N. B. 1929. *Economic Chronicle of the Great War for Great Britain and Ireland, 1914–1919*. London: H. Milford, Oxford University Press, pp. 135 and 140.
23 Journal of the Board of Agriculture, March 1918 p. 1476.
24 *Yorkshire Telegraph and Star* 19 April 1918 p. 3.
25 *The Bournemouth Guardian* 1 June 1918 p. 5.
26 HC Deb 04 June 1918 vol 106 cc1387–8.
27 Dewey, 2014, op. cit. p. 61 ref 7 p. 77.
28 Ernle, Lord. n.d. *The Land and Its People; Chapters in Rural Life and History*. London: Hutchinson, p. 104.
29 Armstrong, 1988, op. cit. p. 50.

30 Armstrong, 1988, op. cit. p. 50.

31 Ministry of Agriculture, Fisheries and Food; Department of Agriculture and Fisheries for Scotland. *A Century of Agricultural Statistics; Great Britain 1866–1966*, pp. 70–4.

32 Holmes, H. 2016. A neglected Innovation: The Double-Furrow Plough in Scotland, Its Early Adoption and Use, 1867 to 1880. *Agricultural History Review* 64(1), 54–80.

33 Gray, M. 1984. Farm Workers in North-East Scotland. In *Farm Servants and Labour in Lowland Scotland 1770–1914*, ed. T. M. Devine. Edinburgh: John Donald, p. 22.

34 https://steamploughclub.org.uk/history-of-steam-ploughing/ (accessed 5/1/2024).

35 Brigden, R. 1984. *Ploughs and Ploughing*. Oxford: Shire, p. 25.

36 Whetham, 1978, op. cit. p. 117.

37 Brigden, 1984, op. cit.

38 Journal of the Farmers' Club, January 1916 pp. 1–20 available from LIBRAL https://bahs.org.uk/LIBRAL/index.html (accessed 5/1/2024).

39 Dewey, 2014, op. cit. ref 11 p. 77.

40 Armstrong, 1988, op. cit. p. 159.

41 Ernle, n.d., op. cit. p. 130.

42 Middleton, 1923, op. cit. p. 226.

43 Dewey, 2014, op. cit. p. 61 and 150 ref 9 p. 161.

44 Ernle, n.d., op. cit. p. 103–4 and 130.

45 Brigden, R. 2000. Equipment and Motive Power. In *The Agrarian History of England and Wales*, ed. E. J. T. Collins. Vol VII Cambridge: University Press, pp. 505–513.

46 Dewey, 2014, op. cit. p. 62 and 151 ref 15 and 13.

47 Lloyd George, D. 1934. *War Memoirs of David Lloyd George 1916–1917*. Boston: Little, Brown, p. 1302.

48 Final Report of the Departmental Committee Appointed by the President of the Board of Agriculture and Fisheries to consider the Production of Food in England and Wales, Cd 8095, p. 6.

49 Dewey, 2014 op. cit. p. 62.

50 Dewey, 2014, op. cit. p. 96.

51 Gregory, A. 2008. *The Last Great War: British society and the First World War*. Cambridge: Cambridge University Press.

52 Conacher, H. M. 1926. Increased Food Production. Chapter IX in *Rural Scotland During the War*, Jones et al. London: Humphrey Milford, Oxford University Press. pp. 122–86, p. 166.

53 Dewey, 2014, op. cit. p. 181.

54 Whetham, 1978, op. cit. p. 99.

55 Dewey, 2014, op. cit. p. 92 and ref 4 p. 104.

56 Dewey, 2014, op. cit. p. 152–3.
57 Dearle, 1929, op. cit. p. 120.
58 Dewey, 2014, op. cit. p. 182.
59 Dewey, 2014, op. cit. p. 178.
60 *Liverpool Echo*, 3 March 1917 p. 3.
61 Dewey, 2014, op. cit. p. 153 ref 25 p. 161.
62 Ernle, n.d., op. cit. p. 131.
63 Dewey, 2014, op. cit. p. 62 ref 19 p. 77 and p. 153 refs 21–23 and 26 p. 161.
64 Dewey, 2014, op. cit. p. 61 ref 6 and 13 p. 77.
65 Ernle, n.d., op. cit. p. 143.
66 Middleton, 1923, op. cit. pp. 231–2 and 235.
67 Whetham, 1978, op. cit. p. 107.
68 HC Deb 17 June 1918 vol 107 cc14.
69 Dewey, 2014, op. cit. p. 155/6.
70 Middleton, 1923, op. cit. p. 235.
71 Barnett, 1985, op. cit. p. 199.
72 Dewey, 2014, op. cit. p. 155.
73 Whetham, 1978, op. cit. pp. 106–7.
74 BRO WAR 4 Reports to the Committee 16/2/18 from Reason, C. M. 1982. Agriculture and Authority—Responses to the Food Production Campaign 1917–1918. MSc University of Birmingham, pp. 94–5.
75 Whetham 1978, op. cit. p. 106.
76 *The Times* 13 March 1902.
77 *The Scotsman* 15 October 1902.
78 Dover Tractor Driver Sent to Assizes. *The Dover Express and East Kent News* 6 April 1917 p. 6.
79 Dewey, 2014, op. cit. p. 116.
80 Dewey, P. E. 1979. Government Provision of Farm Labour in England and Wales, 1914–1918 *The Agricultural History Review* XXVIII, 112.
81 Dewey, 2014, op. cit. p. 119 and refs 63 and 65 p. 143.
82 Journal of the Board of Agriculture, December 1917 p. 1004.
83 Dewey, 2014, op. cit. p. 117 and ref 45, and 6 p. 143.
84 WRO 179/259.9.2 Parcel No. 154 Minute of May 10th, 1918, from Reason p. 89.
85 Whetham, 1978, op. cit. p. 117.
86 Brigden, 1984, op. cit.
87 *Chester Chronicle* 10 March pp. 2 and 8.
88 Journal of the Board of Agriculture, May 1917 pp. 239–40.
89 *Cheshire Observer* 10 March 1917 p. 2.
90 Malden, W. J. 1915. Notes on Machinery and the labour Supply. *Journal of the Board of Agriculture* April pp. 33–37.

91 *The Dundee Courier* 15 March 1917 p. 2.
92 Journal of the Board of Agriculture, October 1917 pp. 759–61.
93 Dewey, 2014, op. cit. pp. 183 and 185–6.
94 Barnett, 1985, op. cit. p. 200.
95 Middleton, 1923, op. cit. p. 225.
96 Lloyd George, 1934, op. cit. p. 1303.
97 Dewey, P. E. 1975. Agricultural Labour Supply in England and Wales During the First World War. *Economic History Review* February 28(1) 100–12.

CHAPTER 5

The Importance of Increased Crop Production to Feed the Nation

This chapter shows how Germany's blockade affected the British population from December 1916 to the end of the war and why the role of the police in maintaining public order in the circumstances was crucial. It demonstrates how the diets of the majority of the British population were threatened and together with drastic food price rises and shortages of staple products, these toxic factors created food instability which frightened the masses, initiating government action to feed the nation from increasing levels of home grown food. Furthermore, reports of food riots in Germany from mid-1916 gave warnings of the seriousness of food shortages experienced by the enemy and how the people reacted:

> GERMAN FOOD RIOTS.
> 28 PERSONS REPORTED KILLED AND 270 INJURED.
> Information received here in regard to the rioting which has taken place in Germany in consequence of the high cost of living shows that the demonstrators have been treated with great brutality by the authorities.
> .... There were tumultuous riots in Berlin, in the course of which twenty-five persons are reported to have been killed and some 200 injured, but it is generally believed that the figures given are much lower than the actual casualties. [3]

The above quotation conveys not only huge rioting in Germany due to high food prices with the consequent large number of injuries, but also

M. Fraser, *The British Police and Home Food Production in the Great War*, https://doi.org/10.1007/978-3-031-58743-6_5

the brutality with which demonstrators were treated. Assaults on farmers to access products at source were also made known shortly afterwards:

> 20,000 IN FOOD RIOT.
>
> German Farmers Fight Mob with Pitchforks.
>
> Travellers reaching Berne from Germany give details of an unprecedented food riot which has just occurred in the countryside near Munich. Believing that the farmers had accumulated large stores of foodstuffs and were holding them up in order to maintain the present famine prices, about 20,000 townspeople delivered violent assaults upon various farm buildings and storehouses. Farmers and peasants fought them with pitchforks and other agricultural instruments. Many arrests on both sides were made by the Bavarian police.
>
> A Geneva message says:- According to the Geneva newspaper "La Suisse," the situation of Munich is becoming graver. Landsturm troops were drafted into the city on Saturday, and if further disorders occur the proclamation of a state of siege seems certain. [4]

This type of propaganda aims to show British readers that although the food situation in Britain may deteriorate, the enemy is suffering even more. This shocking second report of even larger food riots has farmers suspected of hoarding food to maintain high food prices and sets the population against farmers. The rioters were said to have travelled some distance to attack farmers, and further attacks were considered likely, resulting in the mobilisation of troops to quell further disruption. These reports aim to show Britain in a more favourable light, as dealing better with potential food insecurity. Similar reports of food riots with a worsening crisis in Germany, Austria-Hungary, Russia, and France [5] were found in local and national British newspapers from mid-1916, before the British food crisis occurred, and throughout 1917 once the crisis hit. This warned the British government of the potential for unrest on the Home Front if measured action was not taken to reassure and feed the population. It also alerted the police to the potential need to control food riots. Indeed, concern about the British food supply and the rise in the costs of food affecting the diets of the working classes had been debated in the House of Commons since February 1915 [6] and is discussed in Chapter 2, p. 26.

This chapter starts by considering the traditional British diet according to income levels. It shows how the majority of the population, who were the working classes, fed themselves from low wages, resulting in the lowest paid living on subsistence diets relying on carbohydrates; this had been their fate since at least the mid-1880s. The chapter then develops the government's response to preserving and maintaining food supplies, including the measures to increase home food production shown in Chapters 3 and 4.

The worsening food crisis in Britain due to enemy blockade of essential food imports and the severe winter weather produced poor harvests of staple food products such as wheat and potatoes (see Chapter 3). Britain imported 80.8% of its wheat, supplying the population with bread for 42 weeks of the year. In addition to reduced wheat imports from late 1916, shipments of sugar, a further important source of energy in the British diet, declined by around one-sixth from pre-war levels [7]. At home the poor weather and government action to control supplies reduced the British potato crop by around 16%, with the Scottish crop reduced by around a half from pre-war levels; potatoes were also an important energy source and cheap food for the masses. The loss or reduced quality of these three products for many months from March 1917 particularly affected the working classes, around 80% of the population [8]. Whereas less than 20% of people over 10 years of age were classified as middle or upper class according to the 1911 census,[1] this small proportion were less likely to be affected by food shortages and price rises as they could afford to buy more expensive items, making the diets of the masses a serious source of concern [9].[2]

The customary diets of the working classes are shown according to income levels in Table 5.1. The centrality of bread, as well as potatoes and sugar in working-class diets occurred over decades [10, 11].

In Table 5.1, we see the consumption per head of the working-class population in 2,500 households across the UK, including Ireland, from 1887 to 1913. Although evidence is sparse as the numbers in each survey is small, data cover both urban and rural areas. The table shows relatively stable average weekly food consumption between 1887 and 1913, apart from a slight increase in fats and a likely but unknown level of increase in potatoes from 1902. Households with incomes over 30s per week demonstrate a marked increase, particularly in the consumption

[1] These 20% included working in either general or local government, defence of the country, a professional occupation, a commercial occupation or those with private means.

[2] The census also included the populations of Scotland and the whole of Ireland, before partition, and are included in the figures.

Table 5.1 Weekly food consumption by working-class families 1887–1913

	Bread (lbs.)	*Potatoes (lbs.)*	*Sugar (ozs.)*	*Fats (ozs.)*	*Meat (lbs.)*	*Milk (pt.)*
Earning under 18s/week	5.5	1.1	11	3.9	0.9	1
Earning 18s–21s/week	6.5	1.6	16	4.4	1.2	1.2
Earning 21s–30s/week	7.2	1.6	14	4.7	1.5	1.2
Earning 30s+/week	7.3	2.4	17	6	1.8	2.2
Average 1887–1901	**6.7**	**1.6**[a]	**14.4**	**5.2**	**1.4**	**1.4**
Average 1902–1913	**6.6**	3	**15.5**	7.6	**1.2**	**1.8**

[a]Oddy and Miller feel this is an under-estimate
From D. J. Oddy (1970) and D. Oddy and D. Miller (eds.) (1976) pp. 220–222 [12, 13]

of higher cost foods such as meat and milk, compared with those on the lowest incomes, showing how food distribution between the income levels had been uneven at least since the late 1880s [14]. Amongst the lowest paid at the start of the war were British agricultural labourers [15]. Whereas average differences in consumption of meat and milk is around double in households with earnings over 30s per week compared with the lowest paid, consumption of bread, sugar and potatoes stand out as being comparatively higher in the lowest income households, revealing their importance in the diets of the lowest paid working classes [16].

Supplies of sugar, all of which were imported, were secured by government from August 1914 by the establishment of a Cabinet Committee on Food Supplies, creating a Royal Commission on Sugar Supplies chaired by Captain Charles Bathurst which took over the sugar trade [17–19]. This heralded government intervention in food supplies, not only by providing for the British population but also for British, French, and Italian troops.

Despite the German submarine campaign of 1915–1916, there was no problem with food availability until the end of 1916 [20, 21], so that income levels dictated food choice. However, concern for the British food supply grew rapidly from the summer of 1916, although consumption according to the income differentials in Table 5.1 remained until the start of 1917.

Table 5.2 Daily nutritional intake of working-class families from surveys of consumption

	Bread (lbs.)	*Potatoes (lbs.)*	*Sugar (ozs.)*	*Fats (ozs.)*	*Meat (lbs.)*	*Milk (pt.)*
Average 1902–1913 from Table 5.1	6.6	3	15.5	7.6	1.2	1.8
	Energy value	*Protein*	*Fat*	*Carbohydrate*		
Nutrition 1902–1913	2,398	71	65	375		

Translation of average weekly food consumption from Table 5.1 into calorific values appears in Table 5.2. It shows that although nutrition from bread consumption was above average in households earning more than 21s per week from 1902 to the eve of the First World War, potato consumption was likely lower in all households, with the amount of sugar eaten being below the norm for families on the lowest incomes.

Working-class nutritional energy values in Table 5.2 are only 75.6% of the national average consumption of 3,171 calories per head of population and only 72.7% recommended by the Inter-Allied Scientific Commission estimate of 3,300 calories per day that the average man required [22]. This demonstrates that the lower to mid-range of working-class calorific intake since at least 1887 was only barely adequate in quantity and quality, with the poorest being underfed [23, 24]. Moreover, the Royal Society's recommended daily intakes of 100 g of protein, 100 g of fat and 500 g of carbohydrate in which carbohydrates form 71.4% of the diet [25], compared with the diets in Table 5.2, show the reliance on around 73.4% from carbohydrates, such as bread and potatoes, the cheapest foods that supplied the most energy [26], while minimising intakes of fats, meat and milk. Indeed, potatoes were recognised as being more calorific than any cereals [27]. These details of diets by 1914 demonstrate the extent of reliance by working-class households on bread and flour, with 7.33 lbs per man eaten per week, compared with 5.07 lbs per man eaten in the UK as a whole. Poorly paid agricultural labourers not uncommonly ate between 10 and 14 lbs of bread per week [28, 29]. This differential was again seen in 1918 [30]. However, the above intake applies only to the average man estimated to weigh between

15 and 16 stones, but women were known to go without food in order to supplement their husband's diet [31–35], increasing their liability to malnutrition.

These dietary trends continued to the end of 1916 despite steep food price inflation; they show the importance of carbohydrates in most working-class diets. Indeed, total calorific intake did not equal or exceed the recommended 3,300 calories until incomes rose above 40s. per week, the upper end of working-class incomes in industries such as train driving, welding, tool making and riveting [36]. With an improved income, the proportion of protein and fats in the diet increased [37].

Further evidence from a survey of 971 working-class households in June 1918 showed that calorie intake had changed little from June 1914; although less sugar was consumed, the consumption of bread, cereals and potatoes had increased [38, 39]. According to Sumner, despite full employment and wage rises including the war bonus, large increases in the price of food negated higher incomes, despite only modest increases throughout the war in other household costs, such as rent and fuel [40]. This report also provided evidence for an element of substitution of one food product for another by poorer people; for example, bread and potatoes were substituted for meat and cheese [41, 42]. Steep food price inflation also led many men to work large amounts of overtime to gain any advantage [43], which they came to resent. However, this timeframe skips over the intervening period of 1916 to the end of 1917, when Britain's food supply was severely compromised by rising prices [44] and availability with a fall in calorific intake of around 9% by late 1917 [45] and an associated escalation in industrial unrest [46].

Not only bread, but the supply of potatoes decreased dramatically in 1916 and 1917, while prices more than doubled. By 1st February 1917 potato prices had risen by 145% in large towns and 118% in smaller towns compared with July 1914 [47] and by 1st May to 162% in large towns and 123% in smaller towns [48]. Table 5.3 shows the availability and price of potatoes which caused disproportionate stress to the working-class population.

In Table 5.3 we can see the importance of potato growing in the UK including Ireland for consumption by the home population. 98% or more were grown on UK farms or allotments, with only 2% or less imported in any year 1914–1918. The importance for home consumption was a consistent trend with less than 5.5% imported since 1910. The quantity of potatoes bought for human consumption decreased by 21% in 1916

Table 5.3 Potatoes grown and human consumption with price/7lbs in Britain 1914–1918

	1914	*1915*	*1916*	*1917*	*1918*
Quantity for human consumption in 000's tons	4245	4207	3351	3348	5241
Grown in UK (including Ireland)	7926	8015	5433	9254	9873
Imported (in 000's tons)	165	108	89	80	50
Average price in d./7 lbs	4.28	4.05	6.08	9	7.2

From Prest, 1954, Table 27, p. 50 [49]

and by 22% in 1917 compared with 1914 while prices rose by around 42% in 1916 and by 110% in 1917.

The shortage or absence of potatoes from the shops and markets occurred during Lord Devonport's term as Food Controller, appointed by Lloyd George on 22nd December 1916 to nationalise the regulation, supply and consumption of food. The 1916 potato crop was below average, the harsh and prolonged winter, starting with very wet conditions October–December combined with labour shortages, prevented much of the harvest from being lifted and steep price rises began. The harsh winter with low temperatures continued January–March, so that when the potato clamps were opened, farmers claimed many were found to be rotted and disease was prevalent. This serious situation, combined with accusations that farmers aimed to profiteer by removing potatoes from sale, was fiercely debated in the House of Commons on 23rd March 1917:

> Mr. W. THORNE (Labour MP for West Ham South) asked how many tons of potatoes have been rendered uneatable through the last severe frost on account of the farmers not having sufficient covering for the potatoes in their potato clamps?
>
> *Captain BATHURST* (Conservative MP for the South of Wiltshire; Parliamentary Secretary to the Ministry of Food) No statistics of the loss of potatoes owing to frost are available, but there is no doubt such loss has been considerable, and it appears from information received at the Ministry of Food that many farmers, in covering in their potato clamps, did not foresee its unusual duration and severity.
>
> Mr. THORNE Is the hon. Gentleman aware that there is a deep-rooted conviction in the minds of a very large number of consumers that the farmers have done this deliberately for the purpose of bringing about a shortage of potatoes, and so raise prices?

> *Captain BATHURST* No. I know farmers pretty well, and I cannot believe they are such fools. If anything of the sort has happened, it involves a greater loss to the farmer than to anyone else concerned. [50]

While members of parliament used the predominant view that farmers were withholding potato stocks anticipating price rises, government fiercely defended farmers' reputations as honest brokers while protecting the Food Controller's actions. But the absence of potatoes led to ugly scenes in many locations, requiring police intervention, seen in Chapters 6–8.

In Scotland the failure of the potato crop was also said to be largely due to bad weather [51]. As potatoes need to be eaten within a year or they perish, storage beyond one year is not feasible [52], making inedible any of the crop left over from the previous year.

However, claims of farmers not bringing their potatoes to market coincided, in December 1916 with Devonport commandeering the entire potato stock and fixing prices from growers for January and February 1917, and by setting maximum retail prices he attempted to control the supply [53]. Although he was well respected for his retail experience, he was naïve by not taking account of the political economy and processes:

> scientific knowledge is absolutely indispensable for good government. A government that did not take into account this kind of analysis, the knowledge of these processes, which did not respect the result of this kind of knowledge, would be bound to fail. Indeed, we see this when, against the rules of evidence and rationality, government controls the grain trade and fixes the maximum price: it acts blindly; it acts against its interests. [54]

The result was that potato growers felt unable to sell at these fixed prices which angered farmers and potatoes disappeared from markets for many months from February 1917 [55, 56]. The availability, pricing and sale of potatoes became a political football:

> POTATO PRICES
>
> DEADLOCK AT ORMSKIRK MARKET.
>
> An extraordinary deadlock prevails at the weekly farm produce market at Ormskirk to-day as the result of fixing prices for the 1916 crop. Farmers refused absolutely to sell at fixed prices of £8 per ton, while dealers would not give more for fear of having to eventually sacrifice the difference. No trade at all was done except in the case of a few trucks sold on commission.

The Government, however, are still commandeering farmers' potatoes in West Lancashire at £8. [57]

The deadlock with the discourse of growers and salesmen refusing to sell at government fixed prices was justified by growers in different areas of Britain. Barnett [58] shows how farmers "flagrantly evaded the maximum prices order by charging up to £2 per ton for 'carting'":

MR. J. THOMPSON, EAST RIGTON, AND POTATO PRICES.
COST OF PRODUCTION.
Mr John Thompson, of East Rigton, Bardsley, who has been appointed a member of the West Riding War Agricultural Committee .... quoted figures showing the cost of producing an acre of potatoes in Yorkshire. Rent and rates would absorb £1 10s., manure and cartage £8 15s., one ton of Scotch seed with carriage £13 10s., planting and spreading manure 10s., lifting and putting in pie £3, artificial manure £3, twice ploughing £1 10s., dragging and harrowing and carting to pie £2, sorting and weighing and delivering to station £1 10s., making a total of £35 10s. That represented the cost per acre from the time the ground was ploughed to the time the potatoes were delivered at the station ... He believed he under-stated rather than over-stated the case. The average yield for 1916 was 5.93 tons per acre, and it was safe to assume that not more than five tons was edible. This would show a clear loss to the grower of £5 15s. per acre on a fair average crop. The potato crop in Scotland amounted in some districts to about three tons per acre, and if farmers in England were unfortunate in having a crop no larger than that there would be a direct loss to the grower of £18 10s. per acre. [59]

These detailed calculations aim to persuade readers of the justification for farmers and growers to increase the prices for potatoes, attempting to remove themselves from accusations of profiteering, when those living in towns may lack the knowledge of what is involved, and the associated costs.

The police made many arrests for flouting government regulations on retail potato prices. Newspapers across Britain described the situation in early February as a "potato famine" [60], which continued into mid-May [61], while questions in the House of Commons on the availability and price of potatoes continued throughout March. Government intervened in mid-February to pacify farmers and to encourage them to continue to increase crop production by guaranteeing minimum prices for crops, including potatoes [62–65]. However, lack of control of distribution across Britain led to industrial areas at a distance from the farms which grew potatoes, experiencing the worst supply at the highest prices. Devonport resigned on 1st June 1917 due to ill health.

The effects on towns and cities of the lack of potatoes are seen in Chapters 6–8, with the near riot by women in Glasgow, also seen are the locations where deputations by Lord Mayors and Provosts and minor civil unrest developed. New potatoes did not arrive in the markets until June 1917 and even then, with a huge price increase. Potato stocks recovered by 1918 with a bumper crop (see Table 5.3) [66], but prices in 1917 remained higher by 68% compared with 1914 [67]. By 1918 the high potato prices caused the working classes to change their diet so that the average working-class man ate less potatoes per week than the UK norm [68].

A different situation occurred with wheat. To maintain supplies, government amassed stocks from November 1914 [69, 70]. As imports became increasingly scarce, the Food Controller fixed prices and supervised flour mills to control home grown supplies for human consumption, the larger mills from 20th April 1917, others between July and November[3] [71, 72]. To preserve flour, the extraction rate[4] of 70% was raised from November 1916 so that by the end of 1916 extraction had risen to 81.4%, the high point of 92.01% was reached in April 1918. To further eke out wheat supplies, other cereals were added to bread starting in January 1917, such as maize, beans and potatoes. By 13th October 1917 bread contained 17.9% of other cereals, rising to a high point of 32.57% by June 1918. [73, 74]. This changed the consistency of bread to something less palatable and less digestible for some people [75].

Steep price rises in most foods caused working-class families in particular, as well as those on lower incomes, such as policemen, to change their diets in favour of cheaper products, although there was considerable regional variation [76].

The importance of bread in the diets of the working classes was also accompanied by steep price rises until mid-September 1917. In July 1914 a 4-lb loaf, the usual size, averaged around 5 ½ d., by Spring 1917 the cost had risen to 1s. (an increase of 218%) and further increases

[3] Lord Rhondda was appointed Food Controller from 19th June 1917.

[4] The wheat extraction rate is the proportion of the grain milled into flour.

were threatened. The Wheat Commission subsidised bread from 17th September 1917, so that the cost of a 4-lb loaf was reduced to 9d. (reducing the increase from July 1914 to 64%) and flour was sold at a corresponding price [77, 78]. This provided stabilisation until after the war and along with the rising extraction rate and addition of other cereals and potatoes, bread remained available and rationing was unnecessary [79, 80]. The price of milk was also fixed [81].

The Board of Agriculture's plough policy (see Chapter 3) sought to increase the availability of access by the working classes to basic foodstuffs by home food production. It used scientific evidence from the Food (War) Committee set up within the Royal Society in 1916[5] to show how the change of land use could help to feed the population:

100 acres of very poor grass fed to cattle	could feed 2–3 people for 1 year
100 acres of medium grass fed to cattle	could feed 12–14 people for 1 year
100 acres of very good grass fed to cattle	could feed 25–40 people for 1 year
100 acres of fodder beet (mangolds)	could feed 35 people for 1 year
100 acres of wheat as bread	could feed 200 people for 1 year
100 acres of potatoes	could feed 400 people for 1 year[6]

These values made a strong case for increasing the amount of arable land to grow crops solely for human consumption, not to feed animals, so that 100 acres of wheat could support 200 people for a year if used to make bread, whereas the same acreage of finest pasture would only feed 40 people on beef or mutton [82, 83], therefore minimising pasture and meat production. The strategy also minimised changes to working-class diets. Potato growing was promoted in gardens, public parks and on allotments as a source of food for the masses. This policy encouraged and, from 1917, enforced cultivation of crops at the expense of meat and livestock products in the national interest, the former requiring increased labour and/or mechanisation.

By May 1917 the plough policy had not been in force for sufficient length of time to provide results and severe food shortages developed in some areas, reported in the War Cabinet of 7th May to be the most severe in the East End of London where unequal distribution had already led to the poorer classes restricting their diets in "the supply of

[5] The Royal Society (1916) *Food Supply of the United Kingdom* Cd. 8421.

[6] These values were adapted from the Committee's findings, see Chapter 3, p. 61.

certain commodities of food, because these were being bought up by the wealthier classes as economical substitutes" [84]. From late 1917 unequal food distribution caused severe shortages of essential items, such as margarine, particularly in rural areas. Shortages created huge and increasing queues for essential items, which the police supervised and controlled from mid-1917, and brought accusations of hoarding and profiteering with increasing industrial unrest [85]. Furthermore, Britain's wheat stocks had dipped below the danger level to the lowest during the war amounting to only seven weeks supply to feed the nation [86]. The consequent urgency of the need to increase home food production of arable products provides the rationale for the advertisement published in newspapers nationwide (see Chapter 1, p. 5).

By June 1917 with daily newspaper reports of shipping losses and escalating food prices and shortages aggravated and intensified by war conditions, industrial unrest grew across Britain and increased in traditionally militant areas such as the South Wales Coalfield. Lloyd George established a commission of enquiry into industrial unrest in June 1917 [87, 88]. The Commissions, each with three commissioners, were established in eight areas[7] and produced separate reports. Findings demonstrated long-standing issues of class struggle for better work and living conditions, intensified by wartime increases in the cost of living which negated wage rises. The gulf between employers, who resisted wage rises as they reduced profits, and workers who saw a decrease in their real wages by wartime inflation, caused conflict and antagonism. In every area the Commission found that the increased price of food was the most important cause of industrial unrest, although food privation was more severe in Germany, seen in the quotations at the start of this chapter. The collated reports recommended stabilisation of food prices in Britain was more important than wage rises [89]. All reports also showed the widespread preoccupation that a few people were profiteering by amassing huge wealth from providing necessities, with fears that this was the cause of cost-of-living increases (see Chapter 8 for this example in Glasgow), although the reports' authors could find little evidence to support this. Publication in the press from the Scottish report particularly showed trading and shipping companies making huge profits (see Chapter 8, the protests by

[7] The 8 areas were: North-Eastern; North-Western; Yorkshire and East Midlands; West Midland; London and South-Eastern; South-Western; Wales and Monmouthshire; and Scotland.

women at the absence of potatoes). One of the reasons for shortages was highlighted as unequal distribution of food across Britain—abundance in some areas and absence in others, causing deep resentment. Price rises in the shops and markets were also blamed on wholesalers and retailers making large profits, although these middlemen felt the accusations unfair. The collated reports called for more careful government control of food supplies [90].

This critical period of 1916–1917 shows why policemen were released into agriculture to help to minimise the food crisis and forms the subject of this book, the police contribution according to location is given in detail in Chapters 6–8. While their skills and experience helped farmers to increase crop production, their presence in the rural areas and on farms also provided some measure of protection for farmers from the potential invasions seen in Germany in the opening quotations of this chapter. They also provided surveillance of farming to ensure WAEC cultivation orders were implemented, and farmers supported food production policies in the public interest, while working with the growing number of temporary substitutes shown in Chapter 3.

By the end of 1918, measures of increased home food production included a glut of potatoes, and with the control of distribution by rationing, food economy through increasing grain extraction and price maintenance, stabilised food supply [91–93].

In conclusion, this chapter has shown the crucial nature of securing and maintaining the British food supply to feed the majority of the population during the last 2 years of the war. It illustrates how the traditional diets of the majority of the population were threatened from late 1916 until the end of the war. Combined with huge food price inflation, and shortages of traditional food products such as potatoes, this led to widespread fear of national starvation, particularly in the working classes—80% of the population. Fears were emphasised by reports in the press of severe food shortages and food riots in Germany, Austria-Hungary, Russia and France.

This chapter demonstrates how government responded by identifying food insecurity as the root cause of growing industrial unrest. From the outbreak of war government measures secured three major food items: wheat, potatoes and sugar, although the loss of potatoes caused unrest in early to mid-1917. Government action prevented mass unrest from food privation and soaring food prices, while helping to maintain the production of war materials at home. The role of the police as first responders by being placed on farms showed the seriousness attached to increasing home food production to maintain the population, as well as police presence lessening the potential for small-scale invasions and demonstrations to protest at suspected food hoarding by farmers and to steal food. The release of policemen onto farms was widely publicised in local newspapers as a good news story (see Chapters 6–8) and also implicitly warned the public of police scrutiny of farming in the public interest.

The next chapter provides the details of the locations from which policemen were lent to agriculture and the differences between England and Wales, under control of the Board of Agriculture and Fisheries and the Food Production Department, and in Scotland where substitution by policemen was from publicity and pleading for help by the Department of National Service. The next chapter also shows the periods of this furlough and how Chief Constables ensured the release of policemen did not jeopardise their conditions of service. Placement of policemen into agriculture in the following chapters allows us to see the conditions under which they were released, and the impending national disaster of food insecurity already seen in other combatant nations, leading to serious rioting and social disturbances.

References

1 Journal of the Board of Agriculture, April 1915–March 1916 https://archive.org/details/dli.ernet.232728/mode/2up (accessed 25/1/2024).
2 Journal of the Board of Agriculture, April 1917–March 1918 https://archive.org/details/dli.ernet.29517/mode/2up (accessed 25/1/2024).
3 *The Westminster Gazette* May 3, 1916, p. 5.
4 *Lancashire Daily Post* July 11, 1916, p. 2.
5 Ziemann, B. 2014. Agrarian society. Chapter 15 In *The Cambridge History of the First World War*, ed. Jay Winter. Volume II. The State. Cambridge University Press, pp. 382–407.
6 HC Deb 11 February 1915 vol 69 cc756–64.
7 Middleton, 1920, p. 10, ref. 17 in Dewey, P. E. 2014. *British Agriculture in the First World War.* London: Routledge. 2nd edition, p. 35.
8 Barnett, M. L. 1985. *British Food Policy During the First World War.* Boston: George Allen and Unwin, p. 3.
9 Census of England and Wales 1911. Summary Tables Cd. 5705 10 Ed 7 and Geo 5 Ch 27.
10 Marwick, A. 2006. *The Deluge: British Society and the First World War.* Basingstoke: Palgrave Macmillan. Reissued Second Edition. p. 61.

11 Barnett, 1985, op. cit. pp. 112.
12 Oddy, D. J. 1970. Working Class Diets in late Nineteenth-Century Britain. *Economic History Review* 23(2), 314–22.
13 Oddy, D. and Miller, D. eds. 1976. *The Making of the Modern British Diet*. London: Croom Helm, pp. 220–2.
14 Dewey, P. E. 1988. Nutrition and Living Standards in Wartime Britain. In *The Upheaval of War: Family, Work and Welfare in Europe, 1914–1918*, eds. R. Wall and J. Winter. Cambridge: Cambridge University Press, pp. 197–220, p. 199.
15 Dewey, P. E. 1984. Military Recruiting and the British Labour Force During the First World War. *The Historical Journal* 27(1), 199–223, p. 206.
16 Barnett, 1985, op. cit. p. xiv.
17 Dewey, 2014, op. cit. p. 24.
18 Journal of the Board of Agriculture, 1915, pp. 517–18.
19 Rew, R. Henry. January 1918. The prospects of the World's Food Supplies After the War. Paper read to the Royal Statistical Society December 18, 1917, *Journal of the Royal Statistical Society* 81(1), 41–74, p. 70.
20 Beveridge, W.H. 1928. *British Food Control*. London: Oxford University Press, p. 1.
21 Dewey, 1988, op. cit. p. 216.
22 Beveridge, 1928, op. cit. pp. 392–3.
23 Dewey, 1988, op. cit. p. 205.
24 Ferguson, M. 1920. The Diets of Labouring Class Families During the Course of the War, *Journal of Hygiene* 18, 409–16.
25 The Royal Society 1916. *The Food Supply of the United Kingdom* PP 1916 Cd 8421, pp. 3–4.
26 Middleton, T. H. 1923. *Food Production in War*. Oxford: The Clarendon Press, p. 162.
27 Beveridge, 1928, op. cit. p. 153.
28 Beveridge, 1928 op. cit. p. 35.
29 Barnett, 1985, op. cit. p. 108.
30 Dewey, 1988, op. cit. p. 208 Table 6.8.
31 see Fraser, M. 2019. *Policing the Home Front 1914-1918: The Control of the British Population at War*. Abingdon: Routledge, p. 110.
32 Spring-Rice, M. 1939. *Working Class Wives*. London: Virago.
33 Pemberton Reeves, M. 1913. *Round About a Pound a Week*. London: Virago.
34 Clarke, P. 1997. *Hope and Glory: Britain 1900-2000*. London: Penguin Books, p. 91.
35 Oren, L. 1974. The Welfare of Women in Labouring Families in England 1850-1950. In *Clio's Consciousness Raised*, eds. M. S. Hartman and L. Banner. New York: Harper and Row, pp. 50–102.

36 Shpayer-Makov, H. 2002. *The Making of a Policeman: A Social History of a Labour Force in Metropolitan London, 1829-1914.* Aldershot: Ashgate.
37 Dewey, 1988, op. cit. p. 200.
38 Dewey, P. E. 1979a. Food Production and Policy in the United Kingdom, 1914–1918. *Transactions of the Royal Historical Society.* 30, pp. 71–89, p. 78 and 1988, op. cit. p. 206 and Tables 6.8 pp. 209–210.
39 Winter, J. M. 1985. *The Great War and the British People.* Basingstoke: Palgrave Macmillan, p. 216.
40 see also Dewey, 1979a, op. cit. p. 78.
41 Board of Trade. 1918. Report of the Committee appointed to enquire into and report upon (i) the actual increase since June 1914, in the cost of living to the working classes and (ii) any counterbalancing factors (apart from increases of wages) which may have arisen under War conditions. PP Cd. 8980.
42 Barnett, 1985, op. cit. p. 152.
43 Winter, 1985, op. cit. p. 238.
44 TNA MAF 60/562.
45 Dewey, 1988, op. cit. pp. 197–220.
46 TNA CAB 23/3/38.
47 *The Board of Trade Labour Gazette*, February 1917, p. 51.
48 *The Board of Trade Labour Gazette*, May 1917, p. 165.
49 Prest, A.R. 1954. *Consumers' Expenditure in the United Kingdom, 1900–1919.* Studies in the national income and expenditure of the United Kingdom No. 3. Cambridge University Press.
50 HC Deb 23 March 1917 vol 91 c2151.
51 Conacher, H.M. 1926. Increased Food Production. Chapter IX. In Jones et al., *Rural Scotland During the War*. London: Humphrey Milford, Oxford University Press, pp. 122–86, p. 171.
52 Beveridge, 1928, op. cit. p. 154.
53 Journal of the Board of Agriculture, January 1917, p. 1006.
54 Foucault, M. 2007. *Michel Foucault: Security, Territory, Population. Lectures at the College de France 1977–78.* ed. M. Senellart. Basingstoke, UK: Palgrave Macmillan, p. 350–1.
55 Beveridge, 1928, op. cit. pp. 32–50.
56 Gregory, A. 2008. *The Last Great War: British Society and the First World War.* Cambridge: Cambridge University Press, pp. 112–51.
57 *Manchester Evening News* 18 January 1917 p. 5.
58 Barnett, 1985, op. cit. p. 118.
59 *Skyrack Courier* 26 January 1917, p. 7.
60 *The North Wales Weekly News* 1 February 1917 p. 8.
61 *Derby Daily Telegraph* 19 May 1917 p. 3.
62 HC Deb 23 February 1917 vol 90 cc1606.

63 Middleton, 1923, op. cit. pp. 203–4.
64 Barnett, 1985, op. cit. p. 118.
65 Whetham, 1978, op. cit. p. 91.
66 Whetham, 1978, op. cit. pp. 110–1.
67 Prest, 1954, op. cit. pp. 49–51.
68 Dewey, 1988, op. cit. p. 208 Table 6.8.
69 Beveridge, 1928, op. cit. pp. 6 and 10–15.
70 Barnett, 1985, op. cit. p. xv.
71 Beveridge, 1928, op. cit. pp. 106–7.
72 Barnett, 1985, op. cit. pp. 109–110.
73 Beveridge, 1928, op. cit. p. 375 Table XXXI.
74 Dewey, 1979a, op. cit. pp. 86–7.
75 *Report on the Digestibility of Breads* published by the Food (War) Committee of the Royal Society. March 1918, in Beveridge, 1928, op. cit. p. 315.
76 Winter, 1985, op. cit. pp. 225 and 229.
77 Beveridge, 1928, op. cit. pp. 108–9.
78 Barnett, 1985, op. cit. pp. 135–6.
79 Dewey, 1988, op. cit. p. 202.
80 Beveridge, 1928, op. cit. p. 55.
81 Beveridge, 1928, op. cit. pp. 280 and 95–6.
82 Dewey, 2014, op. cit. p. 92 ref 5 p. 104.
83 Ernle, Lord. n.d. *The Land and its People; Chapters in Rural Life and History.* London: Hutchinson, p. 133.
84 War Cabinet minutes 7 May 1917 p. 3, TNA CAB 23/2/154-5.
85 Barnett, 1985, pp. 141–3.
86 Barnett, 1985, pp. 105 and 109.
87 Lloyd George, D. 1934. *War Memoirs of David Lloyd George 1916-1917.* Boston: Little, Brown, p. 256.
88 Chance, W. 1917. *Industrial Unrest: The Reports of the Commissioners (July 1917) Collated and Epitomised.* Published for the British Constitutional Association, London https://archive.org/details/industrialunrest00chanrich/page/6/mode/2up?view=theater (accessed 14/1/2024).
89 Barnett, 1985, op. cit. p. 130.
90 Commission of Enquiry into Industrial Unrest, 1917, op. cit.
91 Winter, 1985, op. cit. p. 223.
92 Barnett, 1985, op. cit. p. 156.
93 Dewey, 1979a, op. cit. pp. 85–6.

CHAPTER 6

Policemen in England Helped Farmers from March 1917 to the End of the War

Police help for farmers in England from March 1917

Following the previous chapters which reveal the build-up and impact of the national crisis of food privation from early 1917, with the threat to local people, families and communities, this chapter gives the locations in which policemen were released to help farmers. With government decrees that Britain must become self-sufficient in crops, the police helped by providing much needed competent manpower during a period of maximum pressure to feed the nation. This chapter shows how the arrangements at national level shown in previous chapters played out locally, with the attempts at substitution into agriculture by different groups and the initial problems and delays, also how increases in mechanical aids for farmers became available or the frustrations of non-delivery. Also included are the effects of various food shortages on local populations.

This chapter opens with a table showing the areas from which policemen were released into agriculture in England and Scotland during March/April 1917. It then focuses on the identified areas in England, giving insights into why help was needed in the various locations from rural as well as urban police forces. It includes the metropolis, although not releasing policemen in the first wave, their contribution began for harvest 1917.

M. Fraser, *The British Police and Home Food Production in the Great War*, https://doi.org/10.1007/978-3-031-58743-6_6

To set the scene, each area includes brief information on the identified police force, where it is known, local geography and Board of Agriculture structures to enforce cultivation and supply manpower and other essential requirements to farmers. It shows farmer's concerns over the sudden expectation that they should feed the nation, how government motivated them to increase crop production and the pressures on everyone with ploughing skills to help. Where knowledge exists, the mechanical and horsepower help is included before identifying how local people were affected by food shortages. This builds the picture of the crisis of threatened population starvation in different communities into which Chief Constables released policemen, assessing their skills and agreeing local terms and conditions for release. Requests for police help were in three phases: (1) Spring 1917 for ploughing and planting; (2) Autumn 1917 for harvest and ploughing; and (3) Spring 1918 for ploughing and planting. Release was mostly full-time with two locations encouraging agricultural work in off-duty time.

Locations from which policemen were released into agriculture in Spring 1917

Table 6.1 gives the locations of the first release of policemen into agriculture in March/April 1917, captured from the weekly police journal *The Police Review and Parade Gossip,* local newspapers, archival material and books. In England it shows only 3 county forces (Cheshire, Norfolk and Somerset), the majority were borough forces, demonstrating how farm labour had moved into the towns to work as policemen, as stated by a city counsellor in early March 1917 (see Chapter 1, p. 4).

In Table 6.1 the numbers are approximate; by comparing different sources small discrepancies were found, possibly due to the initial number of policemen identified with ploughing skills compared with those actually placed on farms. In locations showing "?," numbers could not be found. The total in March/April 1917 was around 400 released throughout England and Scotland. Other locations may also have furloughed policemen, but information was not readily accessible. The two major areas, Birmingham and Glasgow, are developed in Chapter 8 as independent case studies, as their circumstances deserve more detail.

Each area in England will now be shown in turn, the next chapter provides a similar analysis for Scotland.

Table 6.1 Locations of police release in March/April, 1917

England		*Scotland*	
	Approximate numbers		*Approximate numbers*
Birkenhead	4	Ayr	3
Birmingham	100	Berwick, Roxburgh & Selkirk	41
Cheshire	25	Coatbridge	4
Colchester	4	Dundee	40
Liverpool	?	Edinburgh	50
Newcastle-upon-Tyne	6	Haddington County	?
Norfolk	?	Forfar	?
Preston	4	Glasgow	90
Somerset County	8	Greenock	?
St Albans	2	Kincardineshire	?
		Montrose	?

Police help with ploughing and planting in England in Spring 1917

Cheshire: On 23 March 1917, along with Colchester and Norfolk, Cheshire was the first police force in England to report that the Chief Constable, Major (later Lt. Colonel) P. Malcolm, was organising the release of policemen into agriculture:

> CHESHIRE – The Chief Constable is making arrangements to release Police Constables who are able to plough for service on farms in proximity to their residence. [3]

The emphasis was on release of policemen who possessed ploughing skills, the urgency dictated that there was no time to train anyone. Malcolm was appointed Chief Constable in 1910 to the county force of 550; in April 1916 he joined the army but was invalided in January 1917, returning to his post as Chief Constable until 1934. On the outbreak of war 30 army and 4 naval reservists joined their units and Malcolm lent 8 ex-soldiers to the military authorities as drill instructors: (losing 7.6% of the force). In August 1914 he received permission to enrol special constables, 500 were recruited [4]. In addition to the county police force, Cheshire contained 10 smaller borough police forces.[1]

Cheshire, approximately 1,000 square miles in North-East England [5], was traditionally pastureland grazing cattle producing milk, cheese, butter, beef, mutton, and bacon, this made ploughing additional pasture highly contentious. From the beginning of the twentieth century the population increased annually, plateauing at 15% in 1911 [6] due to industrial development and urban residential suburbs serving Liverpool, Warrington, Chester, Macclesfield and Crewe, while the rural population remained static, reflecting the decline in the importance of agriculture. Urbanisation created tensions between the needs of urban and rural communities, particularly during 1917/18 [7].

Cheshire War Agricultural Committee (WAC) was unwieldy with 28 members, it included the chairmen of the 10 District Committees,[2] each divided into smaller areas to undertake the 1917 land survey (see Chapter 3, p. 60 for details of the land surveys). Detailed reports showing which land was in need of attention were reported on 23rd February with cultivation orders to each farmer for individual fields, an example is in Table 6.2.

The WAC was also a co-ordinating body for reports from Cheshire Agricultural Society. Previous District reports presented on 15th January

Table 6.2 Report on Mr Oliver's farm on 23rd February 1917

Field No.	*Acreage*	*Present cultivation*	*Recommendation*
227	6.32	Rough grass	Ploughed up at once and sown with oats
102	2.911	After roots and potatoes	Do. Do
89 & 99	11.105	Part ploughed, part pasture	The portion already ploughed to be prepared for green crops and potatoes

[1] The borough police forces were: Birkenhead, Chester City, Congleton, Hyde, Macclesfield, Stalybridge, Stockport, Wallasey, Warrington and Widnes.

[2] Cheshire's District Committees were: Ashton-under-Lyme; Bucklow; Chester; Congleton; Macclesfield; Nantwich; Northwich; Runcorn; Tarvin and Whitchurch; and Wirral.

demonstrated the widespread alarm at the Board's proposals[3] for increase "at once" in crop production, [8] due to the lack of skilled labour:

> It is worse than useless to break up pasture land while considerable areas of arable land are unable to be properly cultivated for lack of skilled labour. [9]

Dissenting voices from Ashton-under-Lyme district committee indicated refusal to comply without increases in and preservation of the existing labour force:

> the proposed Scheme of the Board of Agriculture for the increased production of home grown food cannot be given effect ... unless some steps are at once taken by the Board to ensure the Farmers having the necessary labour, and to safeguard them from having their present staffs further depleted. [10]

Resolutions sent from the WAC to the Board on behalf of the dissenting farmers pleaded for action on their behalf to ensure the army took no further men from agriculture. A lengthy report of the meeting occupied a prominent position in the weekly *Cheshire Observer*, with a report from Wirral Farmers' Club and support for organised women's labour, alerting readers to the serious demands on agriculture to feed the people [11]. WAC meeting venues included Chester Castle which housed the army depot, showed the close links between army, for release of soldiers to help farmers, and agriculture.

When instructions were delivered to the WAC from the Board of Agriculture to reduce their membership in line with the Board's restructuring this was also contentious, but agreed on 26th January when John Emberton, a farmer and cheesemaker, Chair of Nantwich district committee, the largest rural area [12] became Chair, Colonel George Dixon, Chair of Cheshire County Council and County Commandant of Cheshire Volunteer Regiment [13], became Deputy Chair and 6 other members including three chairs of the largest District Committees (Chester, Northwich and Bucklow) were elected. This caused dissent from districts not represented. Requests to increase the membership to include more district committee chairs were rejected by the Board [14].

[3] See Prothero's proposals sent to WACs in Chapter 3, p. 60.

Newspaper reports gave the votes cast to elect the Chair and members [15] and the resulting protests by the Chair of Wirral District Committee that he gained sufficient support to be represented, portraying Wirral as one of the most important farming areas in Cheshire [16], although it had become urbanised with large areas of development and growth in suburbs serving Birkenhead. The districts of Tarvin and Whitchurch, and Macclesfield also lost their seats.

The revised WAEC membership at its first meetings indicate the total unpreparedness to increase crop production with initial confusion over manpower and equipment for farmers. On 26th January the employment of prisoners of war (POWs) was considered following the Board's letter of 16th January 1917; both Nantwich and Northwich requested a batch of 75 with military guards for 2 months, housed by the district committees. Two batches of 30 were similarly requested for Arclid and Sandbach. Districts eligible to receive help were to draw up lists of farms within a radius of around 3 miles from the designated centre, unless other special arrangements for POW accommodation and guarding were made [17, 18]. However, the Board withdrew the scheme within a month, unless accommodation and guarding arrangements were in place, due to national criticisms of excessive batch sizes and the number due for Nantwich was delayed. The police were responsible for POW control, informed by the Home Office before they arrived, Districts were given responsibility to inform the police if POWs misbehaved [19], but Home Office delays in informing the police of POW arrival led to inertia. By mid-March local resistance to German POWs living and working in the area was evident [20] and schemes were abandoned due to insurmountable difficulties, this decision was also welcomed by Liverpool and Lymm [21]. Schemes were reinstituted from 21st April [22] and districts asked to submit renewed proposals in May, along with the Board's ploughing targets for the 1918 harvest. Therefore, no POW help was available for farmers until the autumn of 1917.

The January WAEC meeting also requested 6 steam ploughs to plough pasture, rotated between districts. However, only half the number were supplied by the end of March, one from the Food Production Department and two privately owned, available for loan.

WAEC members were robust in promoting the Board's policies, reprimanding and encouraging. In March some districts excluded from the committee continued to act independently and were rebuked, named and shamed in the *Cheshire Observer*: their role was merely to report to the

WAEC which retained executive powers. Previous WAC members were publicly admonished for telling smallholders to plough indiscriminately without reference to the WAEC "this had gone so far as to be detrimental to the interests of the County Council"; "agreements were being abrogated" [23]. Some farmers were planning to plough fine pasture, and some had not cultivated land scheduled as arable. Some district committees also lacked energy and were advised of their important work [24] with a reminder in May of the need for speed and efficiency in this vital work.

The WAEC reported renewed help with manpower on 23rd February. 250 soldiers, based at Chester Castle, were available for the county, 25 for each district, followed by Arthur Lee's[4] letter to WAECs of 2nd March [25]. His speech in the House of Commons outlined a policy that skilled men must remain on the land, tractors should work by day and night, and churchgoers encouraged to work in agriculture on Sundays [26]. Night ploughing began in mid-March with acetylene headlamps in three shifts of seven hours each, with an hour for maintenance between shifts, estimated to plough nine acres per day for 40 days and nights and "the police offering no objections to these lights, so long as they are properly shaded when travelling along the roadway" [27], to hinder enemy invasion. Similar reports of night ploughing around Britain demonstrated the benefit of tractors estimated at 12 days to plough 42 acres against 56 days by manual ploughing. Churchgoers were asked to forgo Sunday as a day of rest and to use it to help agriculture in a "superhuman effort" [28] encouraged by senior clergymen.

From the beginning of March public interest in home food production was raised further, with intense pressure on farmers to plough and plant crops and pressure on local authorities, businesses and individuals to help farmers by supplying labour and equipment. Food shortages and measures to try to alleviate them aroused much public interest in the WAEC and its work, with errant farmers highlighted, headlines such as "LAGGARD CHESHIRE FARMERS. 40 FARMS PARTLY CULTIVATED. WAR COMMITTEE'S DRASTIC ACTION" painted a striking picture [29].

[4] Conservative MP Arthur Lee was appointed Director-General of the Food Production Department (FPD) from 28th February 1917, a separate department with independent day-to-day working from the Board of Agriculture.

From 10th March, C3 soldiers[5] organised into Agricultural Companies were available until the end of the war and when placed, others were promised. A further 200 men from the Home Defence Force, including 50 ploughmen were also available until 15th April; 187 applications from farmers had been received [30]. Farmer's requests for soldier labour were to be submitted on forms available from police stations and sent via Labour Exchanges forwarded to Chester Castle, but Northwich complained of delays in supplying soldiers, indicating the Labour Exchanges were inefficient. The WAEC wrote to Divisional Headquarters Office citing delays with 18 applications, subsequently requests for soldier ploughmen were centralised directly from the WAEC to the army at Chester Castle. From 1st May National Service volunteers would also be placed by the WAEC [31], removing the placement of agricultural volunteers entirely from the Labour Exchanges. But providing soldiers on farms for a month was insufficient, the WAEC appealed to the army for their retention at least until crops were harvested [32]. On 7th April agreement was reached that soldier ploughmen could be retained until the end of April [33], further extended later to 25th July, except for Category A men [34].

[5] British Army Medical categories in 1914 were:

A. **Able to march, see to shoot, hear well and stand active service conditions.**

> **Al** Fit for dispatching overseas, as regards physical and mental health, and training
> **A2** As Al, except for training
> **A3** Returned Expeditionary Force men, ready except for physical condition
> **A4** Men under 19 who would be Al or A2 when aged 19

B. **Free from serious organic diseases, able to stand service on lines of communication in France, or in garrisons in the tropics.**

> **Bl** Able to march 5 miles, see to shoot with glasses, and hear well
> **B2** Able to walk 5 miles, see and hear sufficiently for ordinary purposes
> **B3** Only suitable for sedentary work

C. **Free from serious organic diseases, able to stand service in garrisons at home.**

> **Cl** Able to march 5 miles, see to shoot with glasses, and hear well
> **C2** Able to walk 5 miles, see and hear sufficiently for ordinary purposes
> **C3** Only suitable for sedentary work

A turning point for farmers was Lloyd George's speech in the House of Commons on 23rd February [35]. This followed an article in *The Daily Mail* the previous day saying farmers "from Kelso to Kent" were practically on strike and needed help with guaranteed minimum prices for staple items and cheap loans [36]. Guaranteed prices for wheat and oats for six years and potatoes for the 1917/18 harvest were made available, following recommendations of the Milner Committee [37] with advice on minimum wages for agricultural labourers; the announcement greatly heartened farmers. Widely publicised in Cheshire [38] the guaranteed prices received much support the following week from influential farmers who visualised the regeneration of rural life [39], anticipating resurgence of farming from the agricultural decline of previous decades to more progressive work on the land, providing security and prosperity:

> To-day the produce of food dominates the public thought, and it is for us to show that we can rise to the occasion by demonstrating our ability to produce fare more abundantly, in this crisis, the necessary food for our people. [40]

The rallying cry depicted as patriotic, included strong support by farmers for the Board's policies to enforce cultivation, hoping for repopulation and regeneration of rural areas and the need to enforce a minimum wage for farm labourers.

From the beginning of March immense energy to plough and plant crops was evident everywhere. Rousing statements encouraged urgency:

> The Food Crisis is very real. What is demanded is a margin of safety, so that no matter what the result of the submarine piracy, Great Britain will not be starved into acceptance of a peace which would prelude to another and more dreadful war. Sir Arthur Lee .... referred to "the extreme urgency of the subject," and then remarked that only six weeks remain in which to do the necessary farm work. [41]

According to the *Chester Chronicle* the turning point had been reached for Cheshire farmers to throw their weight urgently behind increased food production. Simultaneously National Service placed half column advertisements in local newspapers stressing this urgent need to help farmers, as in the advertisement in Fig. 6.1.

Increased manpower and mechanical help in mid-March raised hopes: 180 C3 soldiers and 200 from Home Defence were allocated to farms in

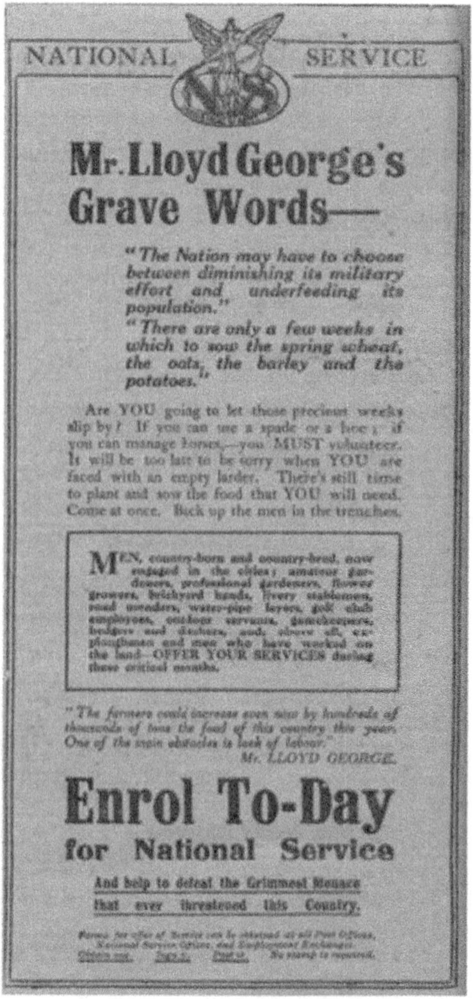

Fig. 6.1 National Service advertisement [42]

Cheshire, including 50 ploughmen, available until 15th April, the WAEC anticipated many applications from farmers. Reports of the number of requests from farmers compared with the numbers of soldiers available became a standing item on WAEC minutes and the numbers on farms were regularly published in Cheshire newspapers. However, Wirral complained that soldiers were only available from 7 a.m. to 5.30 p.m. and until 1 p.m. on Saturdays and passed a motion that the hours should be the same as for other farm workers "an ordinary full day." Also, the 3 promised motor tractors had not arrived and to press their case the committee wrote to Lloyd George saying they had identified 2,000 acres of good quality land to be ploughed but were unable to cultivate it without the retention of soldiers and the supply of motor tractors [43].

Despite the shortage of steam ploughs, the first of 10 motor tractors with ploughs arrived on 3rd March with 2 skilled operators, used initially on grassland which otherwise would remain fallow. The intense cold in February led to more favourable days in March and with the thaw ploughing began with promising reports of farmers planting crops and soldiers starting to work on farms, indicating a significant increase in potato acreage [44] targets for a 1917 harvest were stated as "Back to the Seventies" [45].

By the middle of March three more motor tractors were available under the direction of a district organiser appointed by the Board to maintain and operate them. Motorists were asked to volunteer as tractors drivers and to transport workers to farms. Simultaneously, reports from Arthur Lee's meeting with the Steam Cultivation Development Association said he urged each steam plough should work 7 days a week during daylight hours while he undertook to supply the necessary labour and continuous supplies of coal and water; he appealed to farmers' war spirit that this would make a difference to winning the war [46]. Three more motor tractors were anticipated within six weeks and national reports showed Canada gifting 100 tractors with harrows and binders to Britain, said to be the most efficient model [47]. In addition, the army agreed to lend horses, but only to farms within 12 miles of Chester Castle [48].

However, these upbeat reports coincided with worrying trends in the local availability of some foods. The mayor appealed for self-restraint in food purchases and consumption, saying sensational reports from elsewhere of food shortages, particularly potatoes, did not apply to Cheshire provided everyone moderated their purchases and did not store food, echoing government policy on voluntary food restraint and restrictions on hoarding [49]. However, the following week potato scarcity impacted, occupying many column inches in local newspapers. The mayor made a further appeal for everyone to find food substitutes so that the small number of potatoes could be bought by people unable to afford replacements. National reports of paranoia at the behaviour of farmers hoarding potatoes to create the shortage were reported in the House of Commons by Captain Bathurst, parliamentary secretary to the Ministry of Food Control, who found no evidence of unreasonable withholding of supplies but warned there was an unprecedented shortage of potatoes worldwide and at present consumption, it was likely there would be none by late spring and early summer [50]. His evidence was rejected by the newly formed Fruit and Potato Dealers' Association in Cheshire, comprising local wholesalers and retailers who protested at their dwindling sales, blaming farmers for withholding potatoes for higher prices due to government price fixing. They agreed to a deputation to the mayor "to lay the facts before him" [51, 52].

Alarm at the sudden potato shortage and predicted absence while blaming farmers for withholding stocks to increase prices, threatened the maintenance of law and order in Cheshire. Indeed, Birkenhead saw the police brought in to regulate shoppers besieging the shops for potatoes, angered by rumours of farmers withholding them until prices rose; an inspector from the Food Controller's Department was sent to Liverpool to investigate potato shortages [53]. By early April very few potatoes were available in Chester [54] and from most other retailers in towns. Where found, prices had risen significantly, beyond the reach of many people, they became a luxury item, while claims of plenty were heard from rural areas [55].

Food shortages and hints at national starvation received a high profile in a letter from Lloyd George to local authority mayors, published locally. He called on everyone to help farmers:

> We have reached a crisis in the war when to ensure victory the heroism of our armies at the front must be backed by the self-sacrifice and tireless labour of every one at home ... the production of each quarter of wheat and oats, and each bushel of potatoes is of vital importance. The work of the next few weeks must decide the harvest of the year, and in the nation's interest I urge you, at whatever personal sacrifice, to overcome all obstacles, to throw your fullest energies into the work, and to influence and encourage all who assist you, so that every possible acre shall be sown ... The farmers of this country can defeat the German submarine. [56]

This call by the Prime Minister urged the nation to respond to a national crisis in the war, appealing directly to everyone to show solidarity with the fighting troops, by growing crops to feed the population. The message to everyone from a high-profile figure was to put every effort into growing food, their urgent response would overcome the current crisis and win the war. Simultaneously the WAEC sent a deputation to the Board to plead that no further skilled agricultural workers be called into the army [57]. The impact and publicity of national food shortages, seen locally, also highlighted calls by the National Service Director of Agriculture, J. Harling Turner's appeal for ploughmen to return urgently to the land:

> For the skill which they alone possess there is no substitute, and the time that skill is required is "Now." Many ex-ploughmen are still in the country as gamekeepers, coachmen, or roadmen. Others have migrated to the towns as carters, porters, policemen, etc. They will do national service of the first importance .... If they return to the land for two months or so. [58]

The nationwide appeal to explore all avenues to increase cultivation, led Nantwich Tribunal to exempt farm labourers and ploughmen where labour shortages occurred [59]. National appeals continued the following week with increased urgency; Harling Turner, pleaded for the temporary release of townspeople for work on the land, particularly ploughmen, published in most local and national newspapers:

> Much volunteer help is already in sight. It is not so with ploughing. I therefore appeal to all men who have skill in ploughing to offer their services at once, and to their employers to release them for that work from now until about the end of April. [60]

Readers were reminded that most ploughing was by hand which required skill built up during previous work to be effective. The point being made by the National Service Director of Agriculture in the above quotation is that there was no time to train anyone to plough, the skills needed were to be found in workers who had migrated to other forms of work, often away from agriculture. The urgency to plough and plant for an autumn harvest in the national interest to win the war would be by everyone with latent ploughing skills helping farmers, giving examples of where Harling Turner had located ex-ploughmen—in the police:

> I met a prominent citizen of the provincial town who told me that he had found ten ex-ploughmen in the local Police Force and that he has

> seen the Chief Constable and offered, if these men were made available, to recruit 50 Special Constables in order that these men might be released at once. The Chief Constable undertook to bring this matter under the notice of the Town Council immediately.... I am convinced similar cases can be found all over the country. [61]

This shows public acceptance nationally of substitute part-time special constables for regular police duties during wartime, particularly in the crisis of home food production. Further examples of the release of policemen appeared the following week,

> THE APPEAL FOR PLOUGHMEN In response to Mr Harling Turner's appeal from the National Service Department for men with experience of ploughing and with the warm approval of the county authorities, Mr John Morren Chief Constable of the Counties of Roxburgh, Selkirk and Berwick has decided to lend the services of about forty police constables most of them ploughmen before they joined the service and all experienced in agriculture. The men will as far as possible be employed in their own districts and will be available for their ordinary duties in an emergency. Glasgow has released over eighty of its municipal employees for the ploughing season. Another squad of about 150 will be available in a few days. If the whole of Great Britain were to contribute on the same scale an army of over 10,000 ploughmen would be raised at once and the problem of spring sowing greatly simplified. [62]

Similar claims made by Harling Turner of success in recruiting policemen into ploughing appeared nationwide [63]. Published in a prominent position in newspapers across Britain, the appeal was an urgent national duty, raising the spectre of national starvation if ignored "Every hour we waste of this fine spring weather means less home-grown food in the autumn." By providing comparisons between local authorities, the quotation implored each to examine their workforce skills and to release those with latent abilities in this national emergency; each local authority would be measured against this national standard. The importance of police allocated to farms close to their work, emphasised that they would be recalled to quell any social disturbance, if needed. Locally, the Agricultural Commissioner for National Service for Cheshire and Lancashire, Captain H. Swetenham also pressurised all county, borough, urban and rural district councils to release ex-ploughmen for one month, giving the minimum rates of pay and subsistence allowances [64, 65]. The appeal was followed by the surprise announcement of release of policemen from Cheshire County force:

> The interesting announcement was made that the Chief Constable of the County is making arrangements to release police constables who are able to plough for service on farms in proximity to their residence. [66]

This announcement showed the Chief Constable's assessment of the skills in his workforce to only release those who had suitable experience. On 23rd March the WAEC reported that Chief Constable Malcolm "was prepared to release for fourteen days certain constables experienced in ploughing," also published in the *Runcorn Guardian* (p. 6), district committees would soon receive the men's details, similarly reported in the *Liverpool Echo* and the *Cheshire Observer* on 24th March (p. 6). The *Chester Chronicle* lauded the Chief Constable's actions in a column headed "PATRIOTIC COUNTY POLICEMEN":

> The Chief Constable, Major Malcolm, has notified the Cheshire War Agricultural Committee that there are certain members of the Cheshire Constabulary who before joining the police were engaged on farm work, and have some knowledge of ploughing. Major Malcolm has offered to release these men for a time, provided that they can be given employment within reasonable distance of their own residence ... Consequently the men will be released at once, and steps taken to cover their police duties. [67]

This good news story of release from police duties at once was also reported in the *Liverpool Daily Post* (p. 8). A further report in the *Nantwich Guardian* on 27th March (p. 2) said around 25 constables had been identified and some already placed on farms, confirmed on 31st March in the *Chester Chronicle* (p. 7), warmly welcomed in the Liverpool press [68, 69]. The Chief Constable reported to the Standing Joint Committee on 21st April 1917 [70] receiving approval for policemen and their wives to contribute to war service provided it did not interfere "with the efficiency of the Constabulary," despite 127 men already recruited into army or navy service (25.4% of the force) since the start of the war [71]. As the police were invited onto farms, it allowed increased surveillance of farming practices while performing a crucial role [72]. Chief Constable Malcolm was publicly thanked by the Chair of Cheshire Joint Police Committee "farmers were very thankful for it and for the good these policemen were doing" [73], cementing stronger relations between farmers and the police by acting in the national interest.

The **Birkenhead** Borough police force also communicated with the Cheshire WAEC that they could release four policemen skilled in ploughing [74, 75], but no further information was available.

Preston, the administrative centre of Lancashire, declared police help for ploughing in *The Police Review* on 5th April "Three Preston Policemen, who are skilled ploughmen, are recommended by the Watch Committee for five weeks' work on the land" [76]. However, the Preston Watch Committee recommended on 12th March that although 8 policemen were experienced in agriculture, only 3 had been engaged in this work recently. It was agreed 4 would be released and the difference in pay from the farmer would be made up by the Corporation [77].

J. P. Ker Watson, Chief Constable of the borough police force, was appointed in March 1915 from his post as Chief Constable of Peterborough, to a force of 130. By the end of 1915, 37 were serving in the military (28.5%) [78].

Lancashire's main agricultural areas in the north comprised large estates and small family-run farms in the uplands, particularly in the north-east. But the workforce declined from 36,000 in 1851 to around 20,000 by 1911 as the agricultural depression hit. The southwest also contained arable land with a large potato acreage bearing high yields. The highly industrialised south and midlands specialised in textile manufacture trading internationally, particularly with India, doubling the tonnage of shipping using Liverpool between 1890 and 1914. Other major industries were: engineering, papermaking, rubber and electrical goods, stimulating growth in house building and related trades. Lancashire also led in the manufacture of glass, soap and chemicals in the late Victorian era. The population increased by 1.4% per year between 1891 and 1901 to work in these prosperous industries and in late Victorian times the holiday resorts of Blackpool and Morecambe expanded following the development of rail transport [79, 80]. This shows the sharp increase in Lancashire's industrialisation from mid-nineteenth century while agriculture waned.

The Chairman of the newly formed WAEC, W. Fitzherbert-Brockholes of Claughton Hall near Garstang was part of the "old" gentry officiating in the county council. But the election of WAEC members was equally as troublesome as Cheshire's [81]. Preston District committee represented one of the largest areas, comprising 51 townships [82].

Lancashire mounted a vigorous defence against the army requisitioning 30,000 agricultural workers nationally in February and March 1917 (see Chapter 3, pp. 51). War Office instructions to all army recruiting officers to call up non-exempted men aged 18–22 years, including agricultural workers [83], involved local resistance by the West Lancashire Rural Tribunal, the Board of Agriculture representative and farmers, insisting substitutes must be competent workers before farm hands were removed [84]. The importance of Lancashire's support for home food production is seen in Prothero's[6] visit to Preston on 10th February. He implored farmers to feed the nation including the army, conditional on government help[7] [85], which left farmers worried by his limited power to prevent skilled workers from army recruitment, replaced with unskilled substitutes [86]. Concern for the Board's limited influence was echoed by the Royal Agricultural Society [87].

Notification of 500 soldiers arriving at two centres in Lancashire was announced on 21st February but did not quell the unease [88]. Concerns were raised in the House of Commons by the MP for Leigh, P. W. Raffan, with the reply that around half the farms in Lancashire were staffed below the Bath formula[8] while the remainder only just met the level, the former would receive 15,000 army substitutes from agricultural companies, and

[6] Roland Prothero was appointed President of the Board of Agriculture in December 1916 (see Chapter 3).

[7] Prothero's full address to the Lancashire War Agricultural Committee is at TNA MAF 42/8. HMSO No 163.

[8] The Bath agreement gave the formula:

> The following scale has been agreed between the Army Council and the Board of Agriculture and Fisheries *as a general guide* for determining the number of men to be retained on farms, namely:-
>
> One skilled able-bodied man or lad (wherever possible not of military age) for each of the following:-
>
> each team of horses required to cultivate the land;
> every 20 cows in milk, when the assistance of women or boys is available;
> every 50 head of stall or yard stock, when auxiliary feeding stock is resorted to and the assistance of women or boys is available;
> every 200 sheep, exclusive of lambs, grazed on enclosed lands;
> every 800 sheep running on mountain or hill pasture. (J. Bd. Ag. June 1917 p. 341)

a further 15,000 temporarily from the home defence force. But tribunals remained responsible for deciding release into the army with the Board side-lined, maintaining army superiority in calling up agricultural workers and showing the fierce competition between government departments. Prothero's intervention was unwelcome:

> Mr. RAFFAN: Is the hon. Gentleman aware that the President of the Board of Agriculture has himself stated that is (sic) extremely desirable that this power should be granted to him?
>
> Sir R. WINFREY (Parliamentary Secretary to the Board of Agriculture): Oh, yes; I know that. [89]

Prothero's inability to persuade in Whitehall led him to appoint MP Arthur Lee as Director-General of Food Production, responsible for supplying WAECs with manpower, equipment and supplies. The retention of skilled men on the land was fiercely defended by Arthur Lee in the House of Commons [90], subsequently Lancashire WAEC was notified that 300 soldiers were available until 15th April, 70 were horsemen or ploughmen, and 360 C3 men were available until the end of the war, subject to urgent military requirements. The police held forms at local police stations for farmers to request soldier help. However, farmers were criticised for not making sufficient requests so that some men remained in the depots [91], they were urged by army commanders to submit applications [92].

But using soldier help led Lancashire Farmers' Association to complain that many men sent to farms were incompetent "like square pegs in round holes," that orders changed daily, and different departments issued contradictory instructions. They criticised the tractors sent to Southwest Lancashire as too heavy for the soil, claiming they were a failure, as were some of the men sent to work them—one was a watchmaker! They warned that the serious shortage of skilled labour would hamper food supply [93].

As the urgency to plough and plant accelerated, Preston's Town Clerk received a request from the Military Administrative Officer for Agriculture to release policemen skilled in ploughing. Chief Constable Ker Watson assessed the skills in his force finding 8 ex-ploughmen but only 3 with recent skills, he gained permission from the Watch Committee to furlough 4 for 5 weeks from 12 March [94, 95]. The Town Council agreed to pay the difference between wages from farmers and police pay.

Newcastle-upon-Tyne District War Agricultural Committee was one of 11 districts in Northumberland WAEC, approximately corresponding to rural district areas [96]. William Henry Fitzpatrick Watson-Armstrong (1863–1941), 1st Baron Armstrong of Bamburgh and Cragside, a local benefactor, was appointed uncontested Chair of the Northumberland WAEC. But his unsuitability in business [97] made him a titular chair.

The Chief Constable and Director of the Fire Brigade, Dr J. B. Wright, was responsible for 432 men, including 14 firemen, 15 private constables and 23 in the First Police reserve, a further 40 in the First Police Reserve were allocated to work in munitions. Since the start of the war 31 had left the force as reservists and 85 enlisted: 26.9% of the force, [98] somewhat less than in other forces.

Farming manpower in Northumberland had been depleted by the immense expansion of industry in Newcastle between 1851 and 1914 which drew large numbers of agricultural workers from the neighbouring countryside. The population of Newcastle increased from 87,784 in 1851 to 271,523 by 1914 [99], with vast expansion in the suburbs to house workers in the expanding Armstrong engineering works and the growth of the collieries [100]. Industrial expansion led farmers to increasingly specialise to provide for their nearby markets, further enabled by the development of the railways [101], but change was gradual, rather than the cataclysmic adjustments needed during the First World War. By 1914, supplying the increasing needs of the industrial cities meant farmers in the North were likely the most prosperous in England. The size of farms varied considerably between the 140,000-acre estate of the Duke of Northumberland which maintained many specialised skilled workers, to tiny holdings farmed by the occupier and his family.

Newcastle during the latter years of the war suffered badly from a combination of rising food prices and shortages of staple items, particularly potatoes and potato substitutes. By March 1917 shops had to close early due to the dwindling stocks and the police were often called to control angry crowds. The mayor asked shopkeepers to limit customers to “their usual amounts,” however, the lack of potatoes was said to be a famine and shops further restricted customers to 2 lb each, but supplies were often exhausted before everyone in the long queues were served. The lack of potatoes and fresh fruit and vegetables was portrayed as the cause of an outbreak of scurvy [102]. Desperation by working people saw accusations that farmers were hoarding food, so that by early May the

Cumberland Miners' Association passed a resolution calling on government to instruct the police to search all farms for potatoes and other food stuffs [103].

On 4 May the Chief Constable submitted a letter to the Watch Committee which he had received from the Home Office that requested police help in agriculture. According to Watch Committee minutes, agreement was reached for six constables to be released provided they were paid the same as police pay, and on 30 May and 5 June Constable J. Bell was released for a further period of 14 days [104], showing the PCs acceptability to the farmer. *The Police Review and Parade Gossip* also published a rather different version of the release on 4 May:

> POLICE AND AGRICULTURAL WORK. - Arrangements have been made for the release of a certain number of members of the Newcastle-upon-Tyne Force for work on the land. Two or three men are to be taken from each division for a period of six months. [105]

Only one other location identified individual policemen and the farms on which they worked: **St Albans** in a letter from Constabulary Headquarters: -

> P. C. Bowyer Engaged by Mr G. Wilshin, Kingsbury Farm, St. Albans.
>
> P. C. Russell Engaged by Mr. A. G. Piggott, Cell Barnes Farm, St. Albans.[9]

However, genealogical enquiries through the Police History Society did not find any further information. In all other locations farms were anonymised; also, no indication was found of the type of work policemen undertook, although ploughing was a stated priority. No records exist of how long any remained on the farms or when they returned to routine police duties.

Liverpool newspapers frequently published events and activity in adjacent Cheshire, so that appeals by Harling Turner for the release of local authority employees and his success nationally in recruiting policemen

[9] Town Council of St. Albans Watch Committee, *Meeting of the Watch Committee of the Town Council of the City of St. Albans*—1 May (Town Council of St. Albans, 1 May 1917), SBR_904 p. 209, Heritage Services/Resources Hertfordshire County Council.

also appeared in Liverpool [106] including the release of Cheshire policemen [107, 108].

However, Liverpool Borough police work in agriculture was not included in local newspapers, only in *The Police Review:*

> A number of members of the Liverpool force, who are competent in all kinds of farm work, have offered to devote their spare time – i.e. their annual leave and rest days – to this work. Farmers in the vicinity of this city who wish to avail themselves of their assistance are asked to communicate with the Head Constable. [109]

It is noteworthy that the journal emphasised the competence of the men in farm work, appealing to other readers that these skills were important to ascertain before release into farming. Liverpool was one of two locations[10] where policemen were released in off-duty and holiday time. It coincided with the Mayor of Liverpool's scheme to recruit part-time unskilled labour to help farmers in adjoining districts, receiving interest from National Service. Each volunteer must undertake 6 hours minimum paid work per week [110].

Norfolk. The War Agricultural Committee and later the Executive Committee (WAEC) were chaired unopposed by Sir Ailwyn Fellowes, Conservative and Unionist MP. He was a well-liked local landowner, President of the Board of Agriculture in 1905 until the fall of the government in December that year. From 1917 to 1919 he was Deputy Director of the Food Production Department and Chair of the Agricultural Wages Board. His obituary in September 1924 said he was "the fairest, the wisest, and the most impartial chairman it was possible to find anywhere" [111]. Norfolk WAEC was unusual in its membership of George Edwards, vice President of the National Union of Agricultural and Allied Workers. He reported in January 1917 that farm labourers were not willing to work with German POWs [112]. Initial apathy by farmers continuing into April 1917 and disputes with their farm labourers led to pressure from Ailwyn Fellowes and local newspapers in support of increased and sustained home food production [113, 114]. This included strong pressure on District Agricultural Committees to engage with farmers to send information on their needs for increased production to the WAEC (see Chapter 3, p. 64).

[10] The other location was Norfolk.

Norfolk's main industry was agriculture mainly growing crops, largely wheat and barley, as well as root crops such as turnips and swedes, while livestock was subsidiary [115]. Inhabitants of Norfolk villages were said to dislike newcomers [116] and the separateness of Norfolk labourers is seen in the phrase "Norfolk do different" [117]. Some saw Norfolk as an island, bounded by the North Sea on the East and on the other sides by rivers, with a distinctive building style, a unique dialect and an individual culture [118].

Agriculture was increasingly unionised from around 1890s linked to the Labour party with separation between farmers and labourers, who were increasingly militant. Agricultural labourers in Norfolk were traditionally some of the poorest paid in Britain [119]. From 1916 to 1919 trade union membership rose from around 6,500 to 21,000. With soaring food prices in March 1915, an agreement was reached between the Farm Labourers Union and the Farmers' Federation to increase labourers' wages to 18s. per week. However, when many farmers refused to pay the increase, local strikes broke out involving over 200 men and the Farmers' Federation threatened to use blacklegs. The strike was backed by the Trade Union movement and within a week the increase was paid, the Union claimed an easy victory, although Rew [120] said this rise was not contentious. Farm labourers again struck when soldiers were used to break a strike in June 1915, the bad feeling lasted into July 1916, so that farmers were reluctant to use soldiers and sought other sources of labour. The use of soldiers, women, schoolboys and German prisoners were resisted and, as food prices continued to rise, anti-government feeling developed, particularly when essential food supplies were moved directly to the towns then sent back to the countryside, making availability in the rural areas worse [121]. In February the National Farmers Federation of Norfolk received a deputation from the National Agricultural Labourers' Union demanding 30s. per week. The Federation was willing to pay 25s. if longer hours were worked plus 6d. per hour for overtime [122]. The Union held firm and offered to refer the matter to government arbitration. A further dispute between the National Agricultural Labourers' Union and Norfolk farmers at the refusal to pay the national minimum wage of 25s. set by government in March 1917 was settled by the Board of Trade [123].

The Chief Constable of Norfolk and the Isle of Ely 1915–1927 was Captain J. H. Mander [124]. On 23 March 1917 *The Police Review and Parade Gossip* (p. 91) and local newspapers said the Norfolk WAEC had approached the Standing Joint Committee for police help with ploughing, harrowing and other work on the land. The journal and local newspapers published confirmation that an unspecified number would undertake this work in their off-duty time for a few weeks within their police beat and subject to police duty allowing it. They would receive additional "reasonable remuneration from the farmers for their service" [125]. Due to insularity in Norfolk, it is highly likely these policemen were from the local farming communities making their help well accepted.

Colchester The Chief Constable from February 1913 until 1947 was Captain (later Lt. Col.) H. C. Stockwell, however he re-joined the army in March 1915, not returning to the force until April 1919, leaving E. M. Showers Acting Chief Constable meantime. On 3rd November 1915, following the Derby scheme[11] the Home Secretary wrote to all Chief Constables that every constable who wished to enlist in the army should do so, but "there comes a point beyond which forces could not be depleted without detriment to national interests" [126]. Responsibility for deciding when this point had been reached was with the police authorities. By December 1916 the Acting Chief Constable advised that no more men could be released for military service: 21 out of 61 had already joined (34.4%) [127], a greater proportion than many other forces. However, in March 1917 a further 4 men were released temporarily to plough [128].

The Hon. Edward Gerald Strutt, fifth son of Baron Rayleigh of Terling, Essex, was elected Chair of the Essex War Agricultural Committee [129]. During the war he served on the Milner Committee[12] and was adviser to the Board of Agriculture from February 1917 [130] and a Deputy Director of the Food Production Department [131].

Following initial reticence by farmers to request soldier labour [132] Strutt addressed the District War Agricultural Committees and parish representatives on the importance of local agricultural production and the District Committees' role [133], local newspapers published weekly the large number of soldiers requested by farmers and placed on farms during March 1917 [134, 135]. Reports said they were very satisfactory. However, farmers in White Coln protested at the agricultural survey which showed that between 30% and 80% of their land was badly farmed, they said the surveyors had not visited their farms and the survey was very

[11] See Chapter 3, p. 41.

[12] See Chapter 2, p. 27 for the work and recommendations of the Milner Committee.

superficial [136]. Further protests at the work of the WAEC came from A. Blomfield JP, Chair of the Braintree Farmers' Union who refused to help the WAEC when asked and stated publicly his opposition to their work [137].

Farmers' concerns at paying the national minimum wage of 25s. were also reflected in the Essex National Farmers' Union meeting in March 1917 which reported on Norfolk's settlement by Board of Trade intervention [138–140]. It was towards the latter half of March that the four policemen from Colchester were released into agriculture.

Somerset County police. The Chief Constable, Lieutenant-Colonel Herbert Charles Metcalf was appointed in May 1908, he served until 1939 but was released into the army from 1st December 1914 until 29th May 1919. His deputy, T. Brown acted in his absence. At the Standing Joint Committee on 2nd April, agreement was reached that 8 constables should be released to plough for a "reasonable period" [141].

Requests for police help with Autumn ploughing 1917

Despite efforts to help farmers in Spring 1917, renewed efforts were needed to feed the population in 1918. In April 1917 WAEC Chairmen attended a conference in London addressed by Roland Prothero and Arthur Lee which gave ploughing targets for the 1918 harvest. In May each WAEC received these targets from the Department of Food Production: nationally an addition of 5,000,000 acres of grassland must be ploughed and planted with wheat, 1,000,000 acres with barley and oats and 200,000 with potatoes. Revised estimates were issued in a statement by Arthur Lee targeting an additional 2,050,000 acres of permanent grassland to be ploughed divided between counties, circulated to WAECs in July 1917 [142]. Targets for the counties where policemen were released in early 1917 and where further policemen were released in autumn 1917 are shown in Table 6.3.

Cheshire's WAEC agreed to allocate the acreage to its 10 districts in proportion to their areas of grassland; subsequent district meetings overcame resistance by some farmers who refused to plough more grassland, by threatening to use their ploughing implements to plough and seed the land in the national interest, invoicing the farmer [143]. All WAECs should inform the Board of minimum requirements for labour, horses and machinery to meet these targets. A further land survey was funded

Table 6.3 Revised ploughing targets for 1918

County	*Corn area suggested for 1918*[a]	*Increase in corn area over 1916*	*Estimated area of permanent grass to be ploughed*
Cheshire	136,000	50,000	35,000
Norfolk	506,000	95,000	40,000
East Sussex	86,000	35,000	30,000
West Sussex	96,000	30,000	25,000
Kent[b]	216,000	78,000	65,000

[a]Areas are in acres
[b]These were likely policemen from the Metropolitan Force

by the Board, and districts were asked to make immediate arrangements for this. The survey, conducted independently, aimed to counteract claims by reluctant farmers that their land was unsuitable for ploughing.

However, simultaneously the Army Council issued a further call for agricultural workers under the Military Services (Review of Exemptions) Act. Any farm staffed in excess of the Bath formula or shown on the agricultural census returns of November 1916 to have excess staff (see Chapter 3), was to equalise its manpower of men below category A between farms, but women and soldier substitutes were excluded from agricultural staffing numbers. WAECs were to scrutinise all men eligible for army recruitment and to give or withdraw their approval for army service [144] which led to large numbers of applications for exemption or release into the army being processed by WAECs.

Anticipating the need for harvest labour, Cheshire's WAEC applied to Chief Constable Malcolm for further temporary release of policemen to work with tractors and horse-drawn hand ploughs for autumn and spring ploughing [145] but no further loans to farmers were found [146]. Cheshire WAEC's Machinery and Labour Sub-Committee also requested the Metropolitan Police for help [147] but this was declined [148]. Colchester's Chief Constable also declined a further request from the Food Production Department for police help in October 1917 [149]. However, three further areas (East and West Sussex and Kent) received police help for autumn ploughing.

East Sussex WAEC responded to the Food Production Department's scheme to loan army horses and mules accompanied by skilled ploughmen, provided it increased the arable acreage [150]. *The Police Review and Parade Gossip* published:

> POLICE AS PLOUGHMEN. – Sussex farmers have been informed that the Board of Agriculture, with the object of assisting in the breaking up of grass land, propose as an experiment to allocate 100 horses to agricultural counties, the horses being in charge of Policemen, who have been selected as being capable ploughmen ... Farmers accepting help must plough eight acres of new land during the month. [151]

This work was praised locally as one of six counties to receive help, the ploughmen said to be experienced and carefully selected; further applications from farmers were encouraged [152], the following from **West Sussex**

> POLICEMEN FOR THE PLOUGH. Details were given to the West Sussex Farmers' Union last week ... of the scheme of the Board of Agriculture to provide horses and men to expedite arrears of cultivation, and to facilitate the ploughing up of grass land. If the proposed system ... proved satisfactory, The Board was prepared ... to loan out a large number of horses. Policemen who had experience as ploughmen were to be sent with the horses, to be in charge of them and to plough with them. The first horses were to be sent to the Horsham district – 40 horses, with 16 policemen in charge of them ... Every farmer who took on a team of horses must undertake to break up eight acres of new land in a month, and if a man had already broken up eight acres, he could use the horses. [153]

This experimental scheme was portrayed as speeding up cultivation and ploughing identified grassland where insufficient farm labour was unable to undertake the work. But it stressed that the farmer must agree to cultivate an additional 8 acres in a month and could use the horses if he had already broken up 8 acres. Further reports agreed farmers who had already broken up 8 acres of land were entitled to a pair of horses with ploughing equipment and one man for a month [154]. The reports gave East and West Sussex contribution. But contact with Sussex Archives[13] found no records of policemen released from Sussex; therefore, it is likely these men were from the Metropolitan Police as the section below shows.

[13] I am grateful to Alice Millard, for her archival searches.

The Metropolitan Police were not released into agriculture in the first wave of substitution in Spring 1917; the Commissioner resisted additional activities likely to interfere with police performance in their core work [155], although the police journal showed they could undertake additional work up to 4 hours per day provided it didn't "impair their health and efficiency" [156]. But the Superintendent's Conference the following week rescinded work in munitions in favour of agriculture or horticulture, provided the total working day was less than 11 hours, but opportunities in highly urbanised areas were unlikely [157]. However, a census of men from agriculture was taken in March 1917 to support National Service [158].

In August 1917 Metropolitan policemen who worked on farms before joining the police were allocated to farms in Kent with horses drawn from the military, as part of Arthur Lee's scheme that supplied farmers in Sussex, despite scepticism: "some may criticise the idea of men from the Police Force turning ploughmen" [159]. From Table 7 we see that Kent had a larger area of grass to be ploughed than other counties where policemen were released. The release was during their fortnight's summer holidays [160]. Emsley shows from 1840 to 1930 24.2% of the Metropolitan Police originated from counties bordering the metropolis including Kent [161] which also likely accounts for the police allocated to Sussex in the above quotation.

Although police help to farmers around England was not recognised by government in early 1917, help from the Metropolitan Police was acknowledged by both Lloyd George [162] who said 120 policemen were sent to Kent, and by Prothero:

> Another was the unexpected and valuable windfall of 150 members of the Metropolitan Police, who volunteered their services in their old industry of ploughmen, and ... were employed on the land with excellent results. [163]

The release of metropolitan police to plough was evidently a windfall. The Commissioner of Police, Sir Edward Henry's Annual Report for 1917 also noted:

> 1 Inspector, 6 Sergeants, and 137 Constables were employed by the Food Production Department of the Board of Agriculture and Fisheries. [164]

This help was published in the *Dover Express* on 10 August (p. 8) and was widely welcomed throughout Kent [165] and more widely across England [166].

Although the scheme was widely publicised, the Metropolitan police were not released into other counties across England at a greater distance from London, although they requested reduced price train fares for themselves and their families to travel to Devon and Cornwall for 2 weeks to help with harvest, which was rejected in the House of Commons [167].

Requests for police help with Spring 1918 ploughing and planting and Autumn harvesting

In Spring 1918, with intense trawl by the army for men up to 51 years of age under the Military Service Act (No. 2), exemption certificates were removed from all farm workers aged 18–31 to counter Germany's Spring Offensive, each county received a quota for release. To safeguard ploughing and planting for the 1918 harvest, Cheshire's WAEC asked the Chief Constable of Chester City, J. H. Laybourne, to loan policemen, but was denied [168]. One further request for police help with autumn harvest was found from Surrey Police, where the Chief Inspector's Report of September 1918 shows National Service asked for help with gathering the harvest which was of national importance. The request was said to be dealt with promptly and the Chief Constable, Captain M. L. Sant, received "gratifying reports of the way in which the police came to the aid of the farmers" [169].

No other requests for policemen were found, although in his final report on 30 June 1918, Arthur Lee, Director of Food Production, showed policemen still on the land, likely from Birmingham[170].[14] A section from page 6 of his report is given in Fig. 6.2.

However, by the time of this report, police numbers in agriculture were greatly overtaken by other substitutes, showing the police mainly acted as first responders in the initial crisis of Spring 1917.

The next chapter similarly provides the results of the release of policemen into agriculture in Scotland in 1917–1918.

[14] Although the central column in the report shows the end of year, Arthur Lee resigned in June 1918; further updates were unlikely afterwards. Birmingham police were released from Spring 1917 to the end of the war, see Chapter 8.

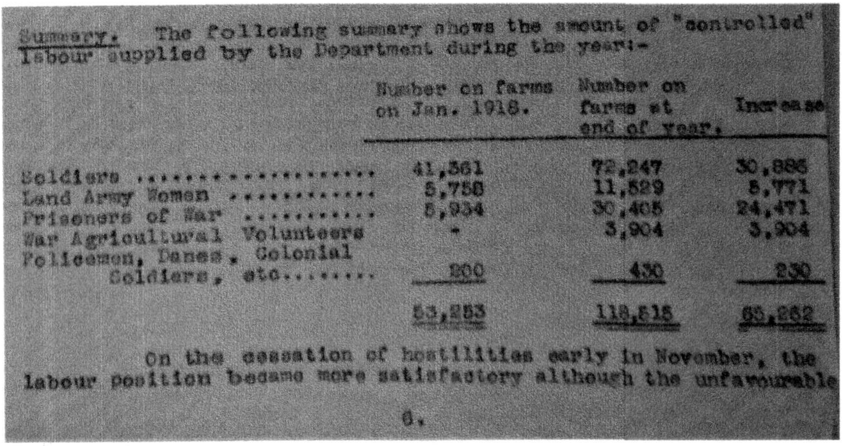

Summary. The following summary shows the amount of "controlled" labour supplied by the Department during the year:-

	Number on farms on Jan. 1918.	Number on farms at end of year.	Increase
Soldiers	41,361	72,247	30,886
Land Army Women	5,758	11,529	5,771
Prisoners of War	5,934	30,405	24,471
War Agricultural Volunteers	-	3,904	3,904
Policemen, Danes, Colonial Soldiers, etc.........	200	430	230
	53,253	118,515	65,262

On the cessation of hostilities early in November, the labour position became more satisfactory although the unfavourable

6.

Fig. 6.2 Report of the food production department for 1918

References

1 Journal of the Board of Agriculture, April 1915–March 1916 https://archive.org/details/dli.ernet.232728/mode/2up (accessed 25/1/2024).

2 Journal of the Board of Agriculture, April 1917–March 1918 https://archive.org/details/dli.ernet.29517/mode/2up (accessed 25/1/2024).

3 *The Police Review and Parade Gossip* 23 March 1917 p. 91.

4 James, R. W. (compiler). 2005. *To the Best of Our Skill and Knowledge: A Short History of the Cheshire Constabulary 1857–1957*. The Museum of Policing in Cheshire.

5 Sylvester, D. and Nulty, G. eds. 1958. *The Historical Atlas of Cheshire*. Chester: The Cheshire Community Council.

6 Phillips, C. B. and Smith, J. H. 2014. *Lancashire and Cheshire from AD1540*. London: Routledge.

7 Harris, B. E. ed. 1979. *A History of the County of Chester Vol II*. Oxford University Press: The Institute of Historical Research.

8 Journal of the Board of Agriculture, 1917 pp. 1009–12.

9 War Agricultural Committee Minute Book Chester City Archives (CCC1/4/1/1).

10 Chester City Archives (CCC1/4/1/1).

11 *Cheshire Observer* 6 January 1917 p. 3.

12 see https://www.perlego.com/book/2447203/chester-in-the-great-war?utm_source=bing&utm_medium=cpc&utm_campaign=2.2.1.1%20DSAG*%20%7C%20UK%20%26%20IE%20%7C%20ALL&utm_term=UK&utm_content=ALL&msclkid=edfaf896eb6f1c3a5afe9d352e007c46 Chambers,

S. 2015. *Chester in the Great War*. Barnsley: Pen and Sword Military, p. 105 (accessed 14/1/2024).

13 https://www.thegazette.co.uk/London/issue/31316/supplement/5420/data.pdf (accessed 14/1/2024).
14 Cheshire War Agricultural Committee. Minute Book. Report of the County War Agricultural Cttee held at Crewe 23 February 1917 p. 1. CCC1/4/1/1.
15 *Chester Chronicle* 3 March 1917 p. 3.
16 *Chester Chronicle* 3 March p. 3.
17 Journal of the Board of Agriculture, 1917 pp. 1134–5.
18 Chester City Archives CCC1/4/1/1.
19 Journal of the Board of Agriculture, 1917 pp. 1018–9.
20 *Chester Chronicle* 10 March 1917 p. 2.
21 *Liverpool Daily Post* 23 April 1917 p. 4.
22 Chester City Archives CCC1/4/1/1.
23 *Cheshire Observer* 3 March 1917 p. 3.
24 *Cheshire Observer* 3 March 1917 p. 3.
25 Journal of the Board of Agriculture, April 1917 p. 97–9 see Journal of the Board of Agriculture Vol xxiv (1917): Bond J. R. Free Download, Borrow, and Streaming : Internet Archive (accessed 7/1/2024).
26 *Lancashire Evening Post* 28 February 1917 p. 4.
27 *Cheshire Observer* 10 March 1917 p. 2.
28 *Cheshire Observer* 3 March 1917 p. 5.
29 *Liverpool Daily Post and Mercury* 13 March 1917 p. 8.
30 Chester City Archives. Report of a meeting of the Executive Committee 10 March 1917 pp. 6–7. CCC1/4/1/1.
31 *The Birmingham Daily Post* 21 April p. 4.
32 *Chester Chronicle* 3 March 1917 p. 3.
33 *Cheshire Observer* 7 April 1917 p. 2.
34 *Cheshire Observer* 12 May 1917 p. 2.
35 HC Deb 23 February 1917 vol 90 cc1598–1607.
36 *The Daily Mail* 22 February 1917, p. 2.
37 see *The Economic Journal* 1916 26(101) pp 105–113.
38 *Chester Chronicle* 24 February 1917 p. 8; 3 March 1917 p. 8.
39 *Chester Chronicle* 3 March 1917 p. 7.
40 *Chester Chronicle* 3 March 1917 p. 5.
41 *Chester Chronicle* 3 March 1917 p. 8.
42 *Chester Chronicle* half column advertisement 10 March 1917 p. 6.
43 *Chester Chronicle* 17 March p. 3.
44 *Cheshire Observer* 3 March p. 7.
45 Ernle, Lord. n.d. *The Land and Its People; Chapters in Rural Life and History*. London: Hutchinson, p. 106.

46 *Cheshire Observer* 17 March pp. 2 and 7.
47 *Chester Chronicle* 10 March 1917 p. 8.
48 *Chester Chronicle* 7 April 1917 p. 7.
49 *Cheshire Observer* 3 March 1917 p. 6.
50 HC Deb 23 March 1917 vol 91 c2151.
51 *Cheshire Observer* 10 March 1917 pp. 7 and 8.
52 *Chester Chronicle* 10 March 1917 p. 8.
53 *Birkenhead News* 7 March 1917 p. 3.
54 *Chester Chronicle* 7 April 1917 p. 8.
55 *Liverpool Daily Post* 13 April 1917 p. 4.
56 *Chester Chronicle* 10 March 1917 p. 8.
57 *Cheshire Observer* 10 March 1917 p. 3.
58 *Chester Chronicle* 3 March 1917 p. 7.
59 *Chester Chronical* 3 March 1917 p. 8.
60 *Chester Chronicle* 10 March 1917 p. 2.
61 *Cheshire Observer* 10 March 1917 p. 2.
62 *Cheshire Observer* 17 March 1917 p. 3.
63 *Hampshire Advertiser* 17 March 1917 p. 5; *Cheltenham Looker-On* 17 March 1917 p. 6.
64 *Cheshire Observer* 17 March 1917 pp. 2 and 8.
65 *Chester Chronicle* 17 March 1917 p. 8.
66 *Liverpool Daily Post and Mercury* 17 March 1917 p. 6.
67 *Chester Chronicle* 24 March 1917 p. 8.
68 *Liverpool Daily Post* 3 April 1917 p. 3.
69 *Liverpool Echo* 2 April 1917 p. 5.
70 Chester City Archives. County of Chester, Standing Joint Committee Minutes April 1916 to January 1919. CCCJ1/1/10.
71 Chester City Archives, CCCJ 1/1/7–11 pp. 263–6.
72 Digby, A. 2000. The local state, Chapter 24 In *The Agrarian History of England and Wales*, ed. J. Thirsk. VII Part II 1850–1914. Cambridge University Press, pp. 1457–64.
73 *Chester Chronicle* 28 April 1917 p. 7.
74 Thompson, S. P. 1958. *Maintaining the Queen's Peace: A Short History of the Birkenhead Borough Police*. Birkenhead. p. 46.
75 *Chester Chronicle* 31 March 1917 p. 3.
76 *The Police Review* 5 April 1917 p. 106.
77 Lancashire Archives CBP50/16.
78 Wilson, S. 2021. *Policing Preston: A history of Preston Borough Police 1815-1969*. Preston: Steven Wilson Publishing.
79 https://www.lancashire.gov.uk/lancashire-insight/population-and-households/population/population-history-1891-2021/ (accessed 14/1/2024).

80 Walton, J. K. 1987. *Lancashire: A Social History 1558–1939.* Manchester University Press, pp. 198–220.
81 *Lancashire Evening Post* 12 February 1917 p. 2.
82 *Lancashire Evening Post* 26 February 1917 p. 3.
83 *Lancashire Evening Post* 9 February 1917 p. 4.
84 *Lancashire Evening Post* 10 February 1917 p. 3.
85 *Lancashire Evening Post* 17 February 1917 p. 4.
86 *Lancashire Evening Post* 22 February 1917 p. 2.
87 *Lancashire Evening Post* 10 February 1917 p. 2.
88 *Lancashire Evening Post* 21 February 1917 p. 2.
89 HC Deb 27 February 1917 vol 90 cc1832–5; *Lancashire Evening Post* 28 February 1917 p. 4.
90 *Lancashire Evening Post* 28 February 1917 p. 4.
91 *Preston Herald* 10 March 1917 p. 7.
92 *Lancashire Evening Post* 17 March 1917 p. 3.
93 *Lancashire Evening Post* 27 March 1917 p. 5.
94 Lancashire Archives CBP50/16 p. 135 (I am grateful to Jacquie Crosby, Archives Services Manager for her searches to provide this information).
95 *Lancashire Evening Post* 23 March 1917 p. 6.
96 *The Newcastle Daily Chronicle* 3 April 1917 p. 2.
97 https://landedfamilies.blogspot.com/2015/08/181-armstrong-later-watson-armstrong-of.html (accessed 14/1/2024).
98 *The Newcastle Daily Journal* 21 April 1917 p. 12.
99 Middlebrook, S. 1950. *Newcastle-upon-Tyne; Its Growth and Achievement.* Newcastle Chronicle and Journal p. 258.
100 Short, B. 2000. Rural demography, 1850–1914. In *The Agrarian History of England and Wales*, ed. E. J. T. Collins. Volume VII 1850–1914 (Part II). CUP, p. 1279.
101 Hallas, C. 2000. The Northern Region. In In *The Agrarian History of England and Wales*, , ed. E. J. T. Collins. Volume VII 1850–1914. Chapter 5E, pp. 403–4.
102 Armstrong, C. 2015. *Newcastle-upon-Tyne in the Great War.* South Yorkshire: Pen and Sword Military, p. 113.
103 *The Newcastle Daily Chronicle* 4 May 1917 p. 4.
104 Tyne and Wear Archives, Newcastle City Council Watch Committee Minutes MD.NC/274/13 pp. 219, 296–7 and 306 (I am grateful to Alyson Pigott in Tyne and Wear Archives for her searches).
105 *The Police Review and Parade Gossip* 4 May 1917 p. 141.
106 *Liverpool Daily Post* 2 March 1917 p. 4.
107 *Liverpool Daily Post* 17 March p. 6.
108 *Liverpool Echo* 17 March p. 2.
109 *The Police Review and Parade Gossip* 20 April 1917 p. 126.

110 *The Liverpool Echo* 25 April 1917 p. 5.
111 Papers of Ailwyn Edward Fellowes, 1st Baron Ailwyn. MC 2001/2. Norfolk Record Office, Norwich.
112 *Norfolk Chronicle* 19 January 1917 p. 3.
113 *Norfolk Chronicle* 12 January pp. 2 and 7.
114 *The Diss Express, and Norfolk and Suffolk Journal*, 1917. 23 March pp. 8 and 20, April p. 8.
115 Norfolk County Council. 1965. *The Official Guide and Handbook to the County of Norfolk*. 4th Edition. Cheltenham and London: ed. J. Burrows; Dutt, W. A. Revised by Long, E. T. 1949 *Norfolk*. London: Methuen.
116 Kennett, David H. 1980. *Norfolk Villages*. London: Robert Hale.
117 Howkins, A. 1985. *Poor Labouring Men: Rural Radicalism in Norfolk 1872–1923*. Routledge and Kegan Paul.
118 Yaxley, D. 1977. *Portrait of Norfolk*. London: Robert Hale.
119 Rew, H. 1923, January 1. The Norfolk Agricultural Dispute. *The Contemporary Review* 123(39), 570–77.
120 Rew, 1923, op. cit. p. 573.
121 Howkins, 1985, op. cit.
122 *Birmingham Gazette* 26 February 1917 p. 5.
123 *Essex Newsman* 24 March 1917 p. 1.
124 Butcher, B.D. 1989. *'A Movable Rambling Police' An Official History of Policing in Norfolk*. King's Lynn: Norfolk Constabulary Headquarters, Norwich.
125 *The Diss Express, and Norfolk and Suffolk Journal* 23 March p. 8; 30 March 1917 p. 102.
126 Stallion, Martin 2012. *Our Duty Has BEEN done: A Record of Colchester Borough Police 1836–1947* p. 153. Braintree: Stallion, p. 152 and Essex Record Office D/B 6 M27/1–14.
127 Stallion 2012, op. cit. p. 153.
128 *The Police Review and Parade Gossip* 23 March 1917 p. 91.
129 Essex Record Office D/Z 45/15.
130 *Pall Mall Gazette* 17 February 1917 p. 5.
131 *The Leeds Mercury* 28 February 1917 p. 3.
132 *The Essex County Chronicle* 2 March 1917 pp. 2 and 4.
133 *The Essex County Chronicle* 9 March 1917 p. 3.
134 *Essex Newsman* 17 March 1917 p. 1.
135 *Chelmsford Chronicle* 23 March 1917 p. 3.
136 *Chelmsford Chronicle* 23 March 1917 p. 3.
137 *The Essex County Chronicle* 9 March 1917 p. 3.
138 *The Essex County Chronicle* 16 March 1917 pp. 3–4.
139 *Essex Newsman* 17 March 1917 p. 3.
140 *Essex Newsman* 24 March 1917 p. 1.

141 *Somerset and West of England Advertiser* 6 April 1917 p. 3.
142 Official Notices and Circulars. The Nation's Food: The Revised Agricultural Programme for 1918. August 1917 pp. 564–6 https://archive.org/details/dli.ernet.29517/page/563/mode/1up?view=theater) (accessed 14/1/2024).
143 *Chester Chronicle* 12 May 1917 p. 3; 26 May pp. 3 and 6.
144 *Cheshire Observer* 12 May 1917 pp. 3 and 8.
145 *Chester Chronicle* 28 July 1917 p. 6.
146 Cheshire War Agricultural Committee. Executive Committee. Minute Book No 2, Cheshire Archives and Study Centre (CCC1/4/1/1); County of Chester. Standing Joint Committee Minutes April 1916 to January 1919. CCCJ1/1/10.
147 Report of a meeting of the Machinery and Labour Sub-Committee 10 August 1917 (p. 8); *Chester Chronicle* 4 August 1917 p. 6.
148 *Chester Chronicle* 4 August 1917 p. 6.
149 Stallion 2012, op. cit. 154.
150 OFFICIAL NOTICES AND CIRCULARS. *J Bd Ag* XXIV, 1, April 1917, 234–5 and 240–1.
151 *The Police Review and Parade Gossip* 27 July 1917 p. 234.
152 *Sussex Express* 27 July 1917 p. 2; *Mid Sussex Times* 31 July 1917 p. 2.
153 Policemen for the Plough. *West Sussex Gazette* 26 July 1917 p. 2.
154 Board of Agriculture's enterprise. *Mid Sussex Times* 24 July 1917 p. 2.
155 TNS MEPO 2/7213.
156 *The Police Review and Parade Gossip* 16 March 1917 p. 81.
157 *The Police Review and Parade Gossip* 23 March 1917 p. 91.
158 *The Police Review and Parade Gossip* 30 March 1917 p. 102.
159 *The Police Review and Parade Gossip* 17 August 1917 p. 262.
160 *The People* Sunday 19 August 1917 p. 5.
161 Emsley, C. 1996. *The English Police: A Political and Social History*. New York: Addison Wesley Longman, p. 198 Table 9.3.
162 Lloyd, George, D. 1934. *War Memoirs of David Lloyd George 1916–1917* Boston: Little, Brown, pp. 234–6.
163 Ernle, *Land* op. cit. p. 144.
164 *Report of the Commissioner of Police of the Metropolis for the Year 1917* [Cd. 9204] p. 3.
165 see for example: London Policemen for Harvest Work. *Kent Messenger and Gravesend Telegraph* 11 August 1917 p. 2; *Maidstone Telegraph* 11 August 1917 p. 2; *South Eastern Gazette* 21 August 1917 p. 2; *Kent Messenger* 18 August 1917 p. 2.
166 for example: *Acton Gazette* 24 August 19197 p. 3; *Whitby Gazette* 24 August 1917 p. 6; *Reading Mercury* 25 August 1917 p. 10; *Illustrated Police News* 30 August 1917 p. 2; London Policemen as Harvesters. *Westminster Gazette* 18 August 1917 p. 8; London P.C.s as Harvesters *The Globe* 18 August 1917 p. 5; *Staffordshire Sentinel* 20 August 1917 p. 2; *Yorkshire Evening Post* 20 August 1917 p. 4; *Western Evening Herald* 20 August 1917 p. 3; Police as Harvesters. *Pall Mall Gazette*. 18 August 1917 p. 5; *Reynold's Newspaper* 19 August 1917 p. 1.

167 HC Deb 07 June 1917 vol 94 c336; HC Deb 21 June 1917 vol 94 c1991W; The Police Review and Parade Gossip. 15 June 1917 p. 189; 6 July 1917 p. 210 and 27 July 1917 p. 234.

168 Minutes of Chester City Watch Committee 11 April and 20 June 1918. ZCCB 22 P 77. Cheshire Archives and Local Studies, Chester.

169 https://www.surreyinthegreatwar.org.uk/story/a-policemans-lot-the-surrey-police-in-wartime-part-4-1918/ (accessed 14/1/2024) (I am grateful to Robert Bartlett for pointing me to this website).

170 TNA MAF 42/8 Report of the Food Production Department for the year 1918.

CHAPTER 7

Policemen in Scotland Helped Farmers from March 1917 to the End of the War

This chapter follows from the previous one by focusing on Scotland's contribution to the release of policemen into agriculture during the second half of the First World War. A separate chapter is necessary, due to the different administrative systems between England and Scotland that managed agriculture and manpower replacement. It builds the picture of the national effort introduced in Chapters 2–5 to overcome the fear of national starvation in which no stone was left unturned to find substitutes to help farmers to plough and plant crops and to provide them with increased mechanical aid to minimise the need for vast increases in manpower. How each area in Scotland providing policemen to plough responded to this crisis is shown below.

Until mid-1916 the loss of agricultural labour from Scottish farms was replenished, although from mid-1915 with losses of 8.2%. The Departmental Committee on Food Production said further development of agriculture was impossible if enlistment continued. But this advice seems to have been consistently ignored, as mid-1916 losses increase to 14.7% with further recommendations that no more farm workers should be enlisted without agreement by the Board of Agriculture for Scotland, by 1917 28.2% had enlisted rising to 36.5% by July 1918, to be replaced by substitutes of less ability [1]. However, these reports do not include the number of substitute workers, such as women and soldiers as are shown

M. Fraser, *The British Police and Home Food Production in the Great War*, https://doi.org/10.1007/978-3-031-58743-6_7

at the beginning of Chapter 3, therefore the net loss from agriculture in Scotland is unclear.

Ploughmen were usually hired workers, hirings of skilled workers such as ploughmen occurred during January to March at hiring fairs, where agreement on wages, benefits such as crops and living arrangements were struck between the farmer and the worker. By 1914 newspaper advertisements for ploughmen vied with this method of hiring ploughmen at fairs [2].

The Board of Agriculture for Scotland based in Edinburgh retained the powers devolved to WAECs in England and Wales (shown in Chapter 3, p. 60) and worked directly with the Commissioner for National Service, Joseph Harling Turner, to find and place soldiers and other volunteers on farms. The Board also supplied farmers with equipment, fertilisers and other requirements. Roland Prothero, President of the Board of Agriculture and Fisheries for England and Wales merely arranged with the War Office for the release of soldiers on furlough and their distribution between England, Scotland and Wales, which passed to Arthur Lee, Director of Food Production in February 1917, but neither had a direct input to Scottish agriculture during Neville Chamberlain's time as Director-General of National Service (December 1916 – August 1917; see Chapter 3, pp. 66–69). *The Times* portrayed the situation:

> It is only in Scotland that the Agricultural Section of the National Service Department deals with military labour. In England it was retained at the request of Mr. Prothero under the control of the Board of Agriculture. [3]

However, on Chamberlain's resignation along with the agricultural commissioners and sub-commissioners in August 1917 over the allocation of soldiers into agriculture, Harling Turner was retained by the Scottish Office in his role as Agricultural Commissioner for Scotland [4].

District or Local Agricultural Committees based on local authority areas were responsible for linking directly with farmers and were accountable to the Board of Agriculture for Scotland. The Board informed them in November 1916 that lists of soldiers experienced in agriculture were available from the army. A Military Substitution Officer for each district was to approach farmers to assess whether they employed any men fit for army service. Where a suitable man was found, they were to work with an Agricultural Representative appointed for the district to agree a substitute with the farmer [5]. However, confusion and contradiction surrounded the substitution of national volunteers into agriculture, seen in a confidential letter as early as 25 January 1917 from the Scottish Office in Whitehall to the Board of Agriculture in Scotland; the Board was told that although Chamberlain had promised around 7,000 C3 men to Scottish farmers in speeches around Scotland, the army were unwilling to release them; this appeared in the national press [6].

Furthermore, by February, although substitute soldiers had been released, their competence was highly criticised by farmers, noted by Professor Wallace of Edinburgh University:

> Of all the lamentable blunders committed by people in authority during the war, there is none more lamentable, because of its disastrous consequences to the vital interests of the country, than the so-called 'substitution of agricultural labour,' and the supply, not of capable agricultural workers from the ranks, but of miscellaneous squads of mainly incompetent men. This was originated on false premises. [7]

This reflected farmers' feelings of being cheated by releasing their skilled workers when most substitutes were found not to possess agricultural skills. The Board of Agriculture acknowledged this lack of efficient substitutes in its December 1916 report [8]. In addition, initially the Board and local committees had difficulty persuading some farmers to apply for soldier labour or found they did not make the best use of them when supplied, so that increases in production were unlikely. Farmers complained that government had not helped them sufficiently to grow the level of food that was expected to feed the nation, and even where food had been grown, harvesting all of it was impossible due to labour shortages. On the other hand, government accused farmers of opposition to taking the substitutes who were sent to them [9].

However, these difficulties were said to be resolved as the year progressed [10] and by April 1917 Stephen Walsh MP for Ince claimed "The Agricultural Section has been able to provide sufficient ploughmen to meet all the ascertained needs of the farmers" [11], but this was evident only in Scotland, reported in the Sixth Report of the Board of Agriculture for Scotland [12]. This claim by Neville Chamberlain in March 1917 created uproar in parliament (see Chapter 3, p. 68).

The Secretary for Scotland confirmed that agricultural work was within the Police Act [13] in March 1917 and could be certified as approved

Table 7.1 Locations of police release in March/April, 1917

England		*Scotland*	
	Approximate numbers		*Approximate numbers*
Birkenhead	4	Ayr	3
Birmingham	100	Berwick, Roxburgh & Selkirk	41
Cheshire	25	Coatbridge	4
Colchester	4	Dundee	40
Liverpool	?	Edinburgh	50
Newcastle-upon-Tyne	6	Haddington County	?
Norfolk	?	Forfar	?
Preston	4	Glasgow	90
Somerset County	8	Greenock	?
St Albans	2	Kincardineshire	?
		Montrose	?

service [14]. The local authority areas that released policemen into agriculture in March/April 1917 in Scotland are shown in Table 7.1 along with the approximate numbers released.

The numbers in Table 7.1 are approximate, as small differences were found in different sources. The Table contains cities with large police forces such as Dundee, Edinburgh and Glasgow and small rural forces such as Ayr, Montrose and Berwick, Roxburgh and Selkirk. However, little is known about the conditions of release or the total workforce in some of these smaller areas. The Scottish local authorities released large numbers,[1] which can be attributed to the work of Harling Turner. The appeal was made, particularly to sceptical and conventional farmers, portraying him as "the man you know", trustworthy and reliable. He wrote to all local authorities in March 1917 pleading for the release of employees with ploughing experience under the National Service scheme. His credibility was high, son of the Factor of the Duke of Portland's estates in Ayrshire, he was commissioner and land agent to the Duke's estates in Ayrshire (22,000 acres) and Caithness (48,000 acres) and a prominent member of the Ayrshire, Kilmarnock and Galston Farming

[1] Both Birmingham and Glasgow will be discussed in the next chapter, as they are outliers and demonstrate special circumstances.

Associations. He was well regarded in his role as a member of Ayrshire County Council and Chairman of Kilmarnock District Committee, a JP for Ayrshire and chairman of Ayrshire local tribunal [15]. As war progressed, he was praised for his work by the Secretary for Scotland:

> In connection with the supply of labour, he could not speak too highly of the work which had been done by Mr Harling Turner, Mr James Adam,[2] and their assistants, to whose skill and devotion the country owed a debt of gratitude. [16]

This demonstrated his national role. He received the Order of the Knight Commander in January 1918 [17].

The circumstances surrounding the release of policemen into agriculture in each Scottish local authority area will now be considered in turn. Scotland contained 31 county and 29 city and burgh police forces, the establishment was 5,953 with Glasgow City police comprising 1,996, around one-third of the Scottish total [18]. By the end of 1917, 111 policemen had joined the navy and 1,953 joined the army, 34.67% of the workforce [19]. Although the 11 locations shown in Table 7.1 are only a little over 18% of the total police forces in Scotland, others may well have been involved but their contribution did not reach the press. However, at least two said they were unable to help due to depletion of staff.

Ayr Burgh Police. Ayr is the administrative town for Ayrshire, situated on the West coast of Scotland. With the growth of the railways in 1840, it developed as a holiday destination. By the end of the nineteenth century, the population had grown to around 31,000.

Chief Constable James Loudon released 3 policemen to plough on farms in the vicinity of Ayr in April 1917, noted in *The Police Review and Parade Gossip* [20], also in *The Ayrshire Post*[3] but no further information was available. But the relatively large number of policemen released from a small town shows the success of the call by National Service.

Ayrshire County Police. Ayrshire is a county of 1,142 square miles on the West coast, predominantly agricultural land, growing green and root crops, oats, wheat and barley as well as dairy farming. In 1914 it

[2] James Adam was General Commissioner under National Service, Agricultural Section for Scotland.

[3] I'm grateful to Alasdair Malcolm, member of the Police History Society for sending me this press cutting.

also contained coal mines, engineering works and fabric manufacturing including carpets [21].

The Chief Constable was C.G. Robertson-Glasgow. On 1st December 1917 the authorised strength of the force was 257 with 185 available for service and 700 special constables [22]. Local newspapers employed the discourse of the national interest to encourage release of local authority employees to help farmers, many of whom they said were previous farm servants, claiming inhabitants would tolerate lower standards of public service if it resulted in increased home food production, a similar claim was made in Glasgow. Harling Turner's appeal to the local authority was published in the press to give an example to local businesses to make similar releases "at once" [23]. The Burgh of Ayr Corporation gave permission for all suitable burgh employees to be released for ploughing on 12 March 1917 [24], giving a steer to the county, following the letter from Harling Turner pleading for the release of council employees for agriculture. But the following week Councillors reported that farmers paid such low wages that nobody would undertake the work [25]. This counteracted the half column advertisement by National Service addressed to employers and the local authority, appealing to their better nature for help in this time of national crisis:

> There are thousands of men in civilian employment who understand ploughing. They are to be found in all occupations recruited from the country; and especially among policemen, carters, dustmen, porters and so on ... Local Authorities and other employers – the country looks to you for help in this critical hour .... **Lend these Men to the State while the ploughing season lasts**. (Bold and underlining supplied) [26]

The plea encourages local authorities to provide the lead to businesses in this time of impending disaster. Further pressure was exerted the following week by publication of a full column letter from Neville Chamberlain as part of National Service Week to boost recruitment of national service volunteers. It appealed to patriotism by civilians, offering the national minimum wage of 25s. and subsistence allowance of 17s 6d per week if working away from home and was published nationwide [27]. On 20th April 1917 the Chief Constable reported that 11 members of the force had volunteered for ploughing, but were not required, only 1 "is giving part of his time to help in the work on a neighbouring farm whenever he is not definitely required for Police work" [28]. This implies that

10 of the police volunteers were screened as not possessing the required skills.

The response to Harling Turner's call for help with agriculture resulted in more unskilled people applying than he could accommodate, requiring him to publish an apology:

> Some of those who offered themselves to the National Service Department for agricultural work have been disappointed that they have not been immediately placed. The explanation is that, at present, the demand is for ploughmen only, and not until later in the year can scope be found for volunteers who are less skilled. [29]

The mismatch resulted in angry letters to the press nationwide that calls for help were excessive. However, agreement was given by Ayrshire Standing Joint Committee that constables would be allowed to undertake this work, although only one was found suitable.

On 11th January 1918 the Chief Constable again informed the Standing Joint Committee that he had received a request from National Service for the release of policemen to plough and that he "hoped to be able to provide a few men" and reminded the committee they had already given permission for this work [30]. It is not known how many were released into agriculture at this time.

Berwick, Roxburgh and Selkirk County Police. The Borders between England and Scotland was a highly rural area with sheep on the hills, crops in the fields and cattle in both [31]. The 1911 census gave the population of Berwickshire was 29,642 occupying 457 square miles. Roxburghshire occupied 666 square miles with a population of 47,192: Selkirkshire's population was 24,601 occupying 267 square miles.

In his annual report ending December 1916, Chief Constable John Morren, responsible for the three county areas from 1909 to 1933, showed his establishment for Berwickshire County police was 28, with an average age of 43 years, by 1918 this had risen to 44 years [32], their average length of service was 19 years [33]. The force was supported by 215 special constables [34]. Selkirk County police had an authorised strength of 15 with 50 special constables appointed shortly after the outbreak of war [35], while Roxburgh's establishment was 40, with an average age of 40.5 years and average length of service of 18.5 years [36], they appointed 127 special constables shortly after the start of the war [37]. The total county establishment was 83, however, 5 joined the armed

Table 7.2 Berwick, Roxburgh and Selkirk County police

	Establishment	*Numbers recruited into the armed forces*	*Numbers available for service in 1917*	*Numbers released into agriculture*	*% of those available for service*	*Number of special constables*
Berwick	28	5	23			215
Roxburgh	40	10	30			127
Selkirk	15	5	10			50
Totals	83	20	63	40	>63%	392

forces from Berwick, 10 from Roxburgh and 5 from Selkirk making the total available for service in March 1917 63, the establishment breakdown is provided in Table 7.2.

The force contained a different personnel profile than most in Britain: the men were of higher-than-average age compared with forces in England and Wales, shown by Emsley [38] to be mostly comprised of men of military age (19–41 until late 1918). Around 50% of the Berwick, Roxburgh and Selkirk force originated from agriculture whereas the Metropolitan police and Leeds Borough police contained a similar high proportion in 1880, which was declining by 1914, to be replaced by semi-skilled workers such as bakers, butchers, bricklayers, carpenters and joiners [39]. Glasgow similarly employed only around one-third from agricultural backgrounds by 1914 despite many who had relocated from the highlands [40]. The long length of service is also surprising, as attrition rates were high elsewhere. The proportion with agricultural backgrounds also points to local recruitment, although Smale [41] shows a slightly larger number were recruited from Aberdeenshire between 1861 and 1883.

The county force was heralded by Harling Turner as a national example of how the police could help agriculture, particularly as it was a sparsely populated rural area:

> If the whole of Great Britain were to contribute on the same scale as this Border police area, an army of over 10,000 ploughmen would be raised at once, and the problem of spring sowing greatly simplified. Special constables who can do extra duty for the next few months will make it easier for ex-ploughmen in the regular force to be temporarily released for work on the land. [42]

This statement was published nationwide with 40 policemen said to be released. If accurate, with a total of only 63 available for service, the proportion released into agriculture was in excess of a staggering 63% of the force, most said to be experienced ploughmen. Harling Turner sent the arrangements for this release to other local authorities and police forces, as an example for their consideration.

However, labour unrest abounded from both farmers and farm labourers so that requests for police release followed disruption by farm labourers in Berwickshire and part of Roxburghshire for increased wages of 35s per week, 10s more than was offered as a minimum wage by National Service. The Scottish Farm Servants' Union had urged men to hold out for a wage of 35s. per week since October 1916, with wage increases seen throughout the year [43, 44]. But despite increases, reports of farm servants' unwillingness to work were seen Scotland-wide and union membership increased. Furthermore, tensions occurred in the Borders between the Union and landowners such as the Duke of Roxburghe, cousin of Winston Churchill, a large landowner and influential on Roxburgh County Council, who antagonised the farm servants' union when he opposed their membership of the local authority joint committee. Tensions between farmers and farm workers in Berwick and Roxburgh were also identified in the House of Commons during debates on the National Service Bill, with questions from William Pringle, Liberal MP for Lanarkshire North West and J.H. Thomas, Labour MP for Derby referring to substitutes into agriculture being potentially used as strike-breakers in a system of industrial conscription, where agricultural labourers refused to be hired for less than 35s per week. If implemented in its proposed form, they argued that National Service could see employees working for less than the minimum wage while being expected to teach substitutes, who received at least the minimum wage, how to work in farming [45]. The Scottish Departmental Committee on Food Production recommended meetings in all parishes between farmers and farm servants to discuss labourers' grievances including wages, hours and conditions of work. Meetings were established in every parish throughout Berwickshire at the end of January to set wages [46, 47]. However, no reports of strikes by farm workers were seen in the press up to the end of April 1917 and by the hirings of farm servants in Berwick and Hawick, maximum wages had risen to 32s per week [48].

Labour shortages, particularly ploughmen, were a continuing problem for farmers, which led to Berwick and Borders Farmers' Association agreeing to resist breaking up more pasture unless assured of sufficient labour [49]. Also, by February, procrastination in meeting farmers' needs for financial security saw appeals to government for urgent action to set minimum prices and to establish financial loans, essential to prevent farmers striking "The farmers are practically on strike from Kelso to Kent" [50]. Lloyd George announced guaranteed minimum prices to farmers for growing corn and potatoes on 23rd February 1917, as an inducement to increase production [51]. This was a turning point in bringing farmers on side.

Amid the disruption, Harling Turner appealed to local authorities and businesses for the skills of ex-ploughmen, their abilities were unique "Ploughmen are urgently needed. For the skill which they alone possess there is no substitute" [52]. They were needed on the land "Now". The appeal was for release until the end of April to do service to the nation "of the first importance", identifying policemen as a source. *The Jedburgh Gazette* published the immediate response sent to Harling Turner by Chief Constable Morren dated 10th March, he had already released two or three policemen, who were currently engaged on farms and a further 30–35 would be made available [53]. Confirmation the following week showed around 40 policemen identified, all said to possess ploughing experience who, with the approval of the county authorities, would be released onto farms in their own districts of Hawick, Appletree, Roberton and Riccarton, to be recalled if needed in an emergency [54, 55]. The Secretary for Scotland agreed the proposal, certified as approved service under the Police Act [56]. The replacement of regular policemen by 120 Special Constables had been in place in Roxburgh since shortly after the outbreak of war [57], however, in March the Chief Constable of Berwick Burgh force was instructed by the Home Office not to allow any special constables to volunteer for National Service, this was not found in other police forces [58].

Hawick District Agricultural Committee was part of Roxburgh. On 16th March it was reported they would break up 235 additional acres of land and that Chief Constable Thom of the Burgh of Hawick had also agreed to loan one police constable to Berryfell to plough. It was agreed to urge farmers and farm servants to work on Sundays [59].

In Chief Constable Morren's annual report for Selkirk, five policemen (half of the force) were employed in agriculture for 198 ½ days (around a month) and all reports by farmers said they performed satisfactorily. The funds received were distributed to cover wear and tear on clothing and to the men who had undertaken additional routine police duties, the remainder was lodged in the police pension fund [60]. The annual reports for Berwick and Roxburgh were not found.

In August further requests for police help were made to Selkirk County police, with agreement to release a further 6 men for harvest work [61]. No reports of further requests from Berwick or Roxburgh County police were found.

Coatbridge Burgh Police. Coatbridge, a town in North Lanarkshire of 6,818 square miles, in the 1911 census contained a population of 43,286, a high proportion of whom were Irish born. It was part of the industrial heartland, a centre for iron works and coal mining but had grim living conditions into the mid-1930s [62].

The Burgh police force employed 47 officers in January 1919 [63], but it is not clear if this included those who had returned from military service. On 19th March 1917 at the Convenors' Committee meeting of the town council, Chief Constable William McDonald offered four policemen with a knowledge of ploughing temporarily for two months, provided they worked close to their districts to be recalled in an emergency. The release was in response to a circular from Harling Turner [64].

Dundee Burgh Police. Dundee, a large town on the East coast of the central lowlands occupies 24 square miles. In 1911 the population was 167,310, mainly engaged in the jute mills and related industries including docks for import and export of goods and materials surrounding the jute trade, and shipbuilding. Of lesser importance was its jam industry, journalism and publishing.

The Chief Constable of Dundee Burgh police from 1909 to 1924 was John Carmichael. He was awarded the King's Police medal in 1924, invested by King George V at Buckingham Palace on 12 February 1925 and appointed Officer of the Order of the British Empire in 1928. His esteem in the role of Chief Constable was shown by the many thousands who lined the streets of Dundee at his funeral in August 1931 [65], although he had retired 7 years earlier. However, no further information about the police force could be obtained from local archives during the period of data collection.

The HM Inspector of Constabulary annual report for 1916 said the force was short of manpower by 22.5% of the total establishment which had been made good by 121,908 special constables with a daily average on duty of 19,793 [66].

On 14th March 1917 the *Dundee Evening Telegraph and Post* (p. 3) announced that 41 members of the Dundee police force had volunteered for agricultural work with National Service, reported to the City Council by Robert Allan, convenor of the police committee. This appeared in *The Police Review and Parade Gossip* on 30th March as a "splendid response" (p. 102). They would be available to farmers for six weeks [67].

However, the difference in police pay compared with wages paid by farmers was controversial. A Town Council meeting report on 10th March showed two members arguing that farmers should pay the full police wage and not expect the council to subsidise their pay, as occurred in other local authorities, naming Glasgow Corporation. In the casting vote, the Provost agreed to the release and, if necessary, the Council would make up the police wages [68]. However, most local authorities claimed the minimum wage of 25s from National Service for substitutes into agriculture plus the subsistence allowance (see Kincardineshire on pp. 171–2 below).

Also published in Dundee the following day was a response by the parliamentary secretary to the Local Government Board, Hayes Fisher, to the assertion by the Liberal Unionist H.J. Mackinder, MP for Glasgow Camlachie that local authorities in Scotland were temporarily releasing their employees to plough; he urged all local authorities in England and Wales to do likewise [69, 70], making unfavourable comparisons.

No further release of policemen into agriculture from Dundee was seen after April 1917.

Edinburgh City Police. Edinburgh is the capital of Scotland, an ancient city situated on the East coast, on the South bank of the river Forth. The population grew rapidly from mid-1800; the 1911 census gave 320,318 inhabitants, an increase of 5.5% since 1901 [71].

The Chief Constable from 1900 to 1935 was Roderick Ross, born to a crofter in Sutherland in 1865. The total establishment for the force could not be found, although 598 were paraded for annual inspection in November 1919 [72]. The beginning of March 1917 saw *The Scotsman* newspaper, based in Edinburgh, publish a three-quarter column report of an interview with Harling Turner on the need to release ex-ploughmen and how local authorities were locating policemen with ploughing skills, he implored all employers to search their personnel records to find ex-ploughmen and consider how many they could release for two months, appealing to the national interest [73]. The following day an editorial claimed any employer not releasing such employees was unpatriotic [74].

Reports that Edinburgh municipal authorities were considering releasing policemen appeared a week later, the authority discussed how differences in the cost of wages was to be handled. The Lord Provost's Committee, responded to a communication from National Service who arranged placement on farms, identifying 50 policemen with agricultural experience and agreed their release for two months; the corporation would make up the difference in pay [75]. 12 were released from each of 4 divisions, this required each night duty rota to be reduced by 4 men and each day rota by 8. Additional remunerated duties would be required from regular officers, and special constables would be asked to amend their hours and to undertake the roles of the regular force. Revised rotas were to be implemented by each Divisional Officer [76]. In 1917 around 93,000 hours of police duty were performed by special constables [77].

However, not all policemen were allocated in the proximity of Edinburgh, as reports in July showed a few had been allocated to Aberdeen, where their arrival on farms was unplanned:

> One tells that he had to tramp eleven miles to a farm only to be told that the vacancy there had been filled. Then he was sent to another farm, where the canny tenant professed his inability to pay the wages arranged under the scheme. He was offered a few shillings per week by a widow on another farm. Other three Constables had similar experiences before settling down.

This unplanned release suggests that Edinburgh police were targeted by the army for retaining policemen from conscription, as 50 were subsequently released into the army in exchange for those returning from agriculture. The policemen sent to Aberdeen also complained of being housed in poor quality accommodation and receiving frugal and monotonous food [78].

Despite the lack of planned work in Aberdeen, Edinburgh police were warmly thanked for their service to agriculture in a letter from James Adam dated 20th October 1917 [79]. But on their return to police duties and under pressure from military representatives, the Chief Constable promised to release 50 policemen into the army [80].

Not only rural areas were cultivated, ploughing Edinburgh parks was underway by mid-March using tractors. One tractor was loaned for up to three weeks including a driver and ploughman [81].

No further requests for police help in agriculture from Edinburgh were found later in 1917 or in 1918.

Haddington County Police. The population of Haddington, 18 miles to the East of Edinburgh, was 3,993 in 1901. It was the retail centre for large grain markets and an agricultural area producing wheat, barley and oilseed rape, occupying a little over 19.17 square miles [82].

Haddington also responded to the National Service request for help with ploughing in Spring 1917. The Chief Constable, S.W. Douglas, reported to the Standing Joint Committee on 11th May that he could see no objection to constables assisting in their districts, as far as their police duties permitted, provided they return home at night. He suggested charging the farmer 5s. per day per constable, the farmer should also provide them with lunch and, if they worked after 5 p.m., also with dinner. He proposed invoicing farmers monthly, disposal of the monies should be decided by the Committee.

On 13th September, Douglas further reported that a few constables had been employed by farmers at intervals and several were currently harvesting in areas within their beat, when police duties permitted [83]. No indication of the numbers released was given, however, because they worked within their beat, they were likely known to the famers and therefore trusted to be able to carry out the work efficiently. It appears that those released worked part-time in agriculture, rather than being released full-time as reported in most other forces.

Forfarshire County Police. The area of Forfarshire was 879 square miles and supported a population of 281,419 in 1911. On the East coast, it was largely rural with arable land growing oats, barley and wheat and pastureland grazing cattle and sheep, the proportion of arable and pasture was approximately equal [84].

The Chief Constable R. T. Birnie attended The Sanding Joint Committee meeting on 9th April 1917 where a letter from Harling Turner was read pleading for County Council employees help with ploughing "The immediate and imperative demand is for ploughmen … it is essential to recall to the land men who are already skilled" which challenged them that, if they responded, it would also encourage private employers to react likewise. The Chief Constable said he had already sent as many constables as could be spared.

Forfar county police were furloughed to farmers on 3 occasions, Spring and harvest 1917 and Spring 1918. On 18th October Birnie reported the police had worked in various districts in the county, implying they were moved between farms, and the funds received from their work in Spring and Autumn had been handed to the County Treasurer, although they had also received police pay while on agricultural furlough. An amount for work in the Spring 1918 was also lodged [85].

Greenock Burgh Police. The population of Greenock was 68,217 in 1911. On the Southern shore of the Firth of Clyde, 22 miles below Glasgow, it developed a large shipbuilding industry with a port and international trade. Industrially it contained large sugar refineries. During the First World War a large torpedo factory employing 700 workers designed and tested these weapons [86].

Chief Constable James Christie took up post in 1913 [87]. On 20th March 1917 *The Daily Record and Mail* advertised for an increase of 200 special constables, so that an unspecified number of regular policemen could be released into agriculture (p. 4). The corporation agreed the Chief Constable should ask members of the force to volunteer for land cultivation and "a number have at once responded" [88]. The corporation magistrates agreed to make up any differences in pay [89].

Kincardineshire on the East coast of Scotland occupied 380 square miles with a population in 1901 of 40,923. It was a rural area comprised mainly agricultural land growing oats and root crops, the other main industry in 1911 was fishing [90].

The Chief Constable was Charles George. A report in *The Stonehaven Journal* [91] said that around half of the force had volunteered; they would work close to their districts and return home at night. Sergeants were not allowed to be released as their police duties would increase correspondingly. On 29th March Kincardineshire Standing Joint Committee [92] acknowledged receipt of a circular from National Service and a memorandum which gave the arrangements made by Berwick, Roxburgh and Selkirk for the release of policemen to plough. A letter from Sheriff J. Campbell Lorimer agreed to the constables' release for ploughing. *The Police Review and Parade Gossip* reported on 30th March (p. 102) concurred that around half the force had volunteered and would be placed close to their police districts to enable them to return to their stations at night.

The Standing Joint Committee also reported on 19th July (p. 129) that the men were rewarded by receiving a bonus of half the money they earned, in addition to their police pay, those who had undertaken additional work to cover for them would also receive an additional sum and the remainder would be transferred to the Police Fund.

A further release for harvest work, reported to the Standing Joint Committee on 20th December showed nine constables were released to help with the harvest under the same conditions as in March. They undertook storing potatoes in addition to other farm work.

On 18th October 1918 Chief Constable George reported the third release for harvest and the funds earned [93].

Montrose Burgh Police. Montrose is a seaport in Forfarshire, north of Dundee occupying 2 miles in length by 0.75 miles wide. The population in 1911 was 10,974 [94]. The main industry was flax-spinning, other employment came from making linen, canvas, starch, soap, chemicals and rope. The fisheries were also an important source of work [95]. In 1913 it housed the first operational military aerodrome in the UK.

The Chief Constable in 1917 was Alexander Marr. The authorised strength of the force was 12 and more than 60 special constables undertook patrol duties day and night [96].

Montrose Town Council held a public meeting on 9th March to raise the profile of National Service and to recruit volunteers [97]. The following week they reported receipt of a circular from National Service requesting local authority help with agriculture. The only employees identified with skills were policemen. The same day a half column advertisement appealing for ploughmen, addressed to employers and local authorities appeared in the local press [98].

Montrose Town Council agreed to help agriculture by occupying under-utilised ground in North Street and to the east of Northesk Road, around 7 acres, for two years, using the compulsory DORA powers (see Chapter 3, p. 62) to plough and plant potatoes. Planting was undertaken by the police along with more than 30 special constables and 8 boys from Dorward's House of Refuge. Planting one of the fields of 3 acres began at 6am when the Provost and honorary Treasurer arrived and was completed by lunchtime [99]. These single source newspaper articles portray the local authority confiscating unproductive land and providing leadership and manpower, including from the police, to cultivate the land in the interests of the local population.

In conclusion, the release of policemen into agriculture in these eleven areas in Scotland shows that both urban and rural police forces were involved in approximately equal numbers, see Table 7.3.

Table 7.3 County or Burgh police forces involved in agriculture in 1917

	County	*Burgh*
Ayr		1
Ayrshire	1	
Berwick, Roxburgh and Selkirk	1	
Coatbridge		1
Dundee		1
Edinburgh		1
Haddington	1	
Forfarshire	1	
Greenock		1
Kincardineshire	1	
Montrose		1

Considerable diversity in the size of the forces is also evident, from the smallest, Montrose, with an establishment of 12 to around 600 in Edinburgh City. The proportion released from most forces is known, however, of the two that released half or more (Kincardineshire and Berwick, Roxburgh and Selkirk) only in the latter was industrial unrest by farm workers evident. Therefore, the proportion released across Scotland did not appear to be related to keeping the peace, the evidence suggests helping to increase production while surveillance of working practices to comply with government priorities was the more likely reason.

Some locations released policemen entirely from their police duties, such as in Coatbridge, substituting with special constables or with extra duties by the remaining police, as in Edinburgh. Other areas agreed to police involvement in agriculture either only when their police duties permitted, implying occasional periods as in Haddington, or for a short, concentrated period for specific activities, such as planting potatoes in Montrose.

On the initial occasion on which local authorities considered releases, in Spring 1917, the main concern was wages paid by the farmer and whether the police were to be compensated by the local authority when wages received from the farmer were lower than police pay. As Dewey [98] shows, substitute soldiers were expensive, as were policemen compared with agricultural labourers' wages. But following government agreement to minimum prices for crops, Dewey found farmers gained the most financially from farming during the war, compared with farm labourers and landowners, the latter lost out as land prices fell. Indeed, in some areas farmers and farm servants were known to be "making big money at present" even before government agreed minimum crop prices [100, 101]. However, most local authorities agreed to supplement police wages, despite protests from some councillors, as in Dundee, that farmers should pay police wages in full. Indeed, some local authorities received in excess of police pay, as in Kincardineshire and Berwick, Selkirk and Roxburgh, likely due to claiming National Service minimum wage and subsistence, so that they were able to give a bonus to those who worked on the land and also to others whose workloads increased accordingly, as well as providing funds for the police pension.

A further consideration was that most forces insisted those released worked close to their police stations and returned home at night, which appeared to be uncontested. Only Edinburgh released policemen to a distant location, implying that their services were dispensable, even in an emergency; borne out by the Chief Constable who was pressurised to release the equivalent number into the army on their return from farms.

However, two of the county forces, Berwick, Selkirk and Roxburgh and Kincardineshire released a far higher proportion of their men than any other in either England or Scotland, this deserves an attempt at explanation. Was this use of police as ploughmen challenging the employment of ploughmen in these counties and possibly in other counties also, as implied in debates in the House of Commons on the National Service Bill in February 1917?

> The proposal in regard to labour is that a man is at least to be guaranteed 25s. a week, but that if the standard rate in the district to which he is transferred exceeds 25s., he will receive that higher rate. But what is to happen if there is a dispute going on in the particular area for which it is desired to recruit labour? I will give an example. In the county of Berwick, and to some extent in the county of Roxburgh, there is a dispute going on in regard to the wages of agricultural labourers. They are not concerned about a paltry 26s. a week, because what these men are demanding is 35s. a week, and they are refusing to hire themselves at the customary hiring. If this Bill passes into law, is the Director of National Service entitled to draft the volunteers whom he succeeds in getting under this scheme into these counties to obtain the employment of these men whom he gets at 25s. in order to break the agitation of the men for higher wages? [102]

This speech appeals to government to reconsider movement of labour from non-essential into priority industries, which, under current proposals, could increase industrial unrest. It continued the review by Auckland Geddes[4] in December 1916, that more than 3 million men were working in industries of little national importance, such as hotels and entertainment, but how to move them into work that supported the war effort was highly contentious, pointing to the threat of strike breaking. However, there is no evidence from the above locations that the majority of regular ploughmen refused to be hired and that police forces were drafted in to replace them. Whereas most local authorities claimed the same level of pay from farmers that the latter paid their workers, there was still a shortage of ploughmen. *The Scottish Farmer* weekly Situations Vacant columns showed around 10% of advertisements were for ploughmen throughout southern Scotland during March, indicating work was plentiful. However, in Berwick, Roxburgh and Selkirk where industrial action by ploughmen was rife, release of a larger proportion of the county force (63%) was evident compared with other areas in Scotland. But in the other location that released a higher proportion, Kincardineshire where around 50% were released, there was no evidence of industrial unrest by farm workers. Therefore, claims of the police being used as strike-breakers are at best unproven.

The next chapter develops the two areas where larger numbers of policemen were released, Birmingham and Glasgow, and in the former for a longer period than in any other force. These two unusual cases also show the local circumstances and political influences at play.

References

1 Jones, D. T. et al. 1926. *Rural Scotland During the War*. London: Humphrey Milford, pp. 202–5.
2 Robson, M. 1984. The Border Farm Worker, Chapter 5 In *Farm Servants and Labour in Lowland Scotland 1770–1914*, ed. T. M. Devine. Edinburgh: John Donaldson, pp. 80 and 93.
3 *The Times* 10 March 1917 p. 6.
4 *The Scotsman* 18 August 1917 p. 7.
5 Board of Agriculture for Scotland, Circular letter to the Secretaries of the District Agricultural Committees in Scotland from H. M. Conacher. 20 November 1916. National Records of Scotland GRO5/1947.

[4] Director of Recruiting for the Army.

6 NAS Scottish Office, Whitehall to H. M. Conacher, Board of Agriculture, Edinburgh 25 January 1917. HH31/28/51.
7 *The Scottish Farmer* 13 October 1917, p. 746.
8 Scottish Departmental Committee on Food Production Second Report (1917) National Records of Scotland. HH31/24/6.
9 *The Weekly Bulletin* 1 March 1917 p. 26.
10 Board of Agriculture for Scotland, Sixth Report for the year ending 31 December 1917. 1918. Cd. 9069 p. 1i.
11 *The Scotsman* 21 April 1917 p. 9.
12 Cd. 9069. 1918. op. cit. p. 1i.
13 Police (Scotland) Act 1857 [20 & 21 Vict. VII].
14 *Jedburgh Gazette* 23 March 1917 p. 3.
15 *The Kilmarnock Herald and North Ayrshire Gazette* 23 March 1917 p. 3.
16 *Perthshire Advertiser* 16 January 1918 p. 2.
17 *Sheffield Daily Telegraph* 8 January 1918 p. 6.
18 Report of the Committee on the Police Service of England, Wales and Scotland Cmd. 253 HMSO 1919 p. 5.
19 Ferguson, A. G. 1918. Sixteenth Report of His Majesty's Inspector of Police for Scotland, March 1918, Cd. 9012.
20 *The Police Review and Parade Gossip* 20 April 1917 p. 126.
21 See https://en.wikisource.org/wiki/1911_Encyclop%C3%A6dia_Britannica/Ayrshire (accessed 14/1/2024).
22 Glasgow City Library, archives SR22/74/5 County Chief Constables Office Report Book No. 10. 1 March 1918 p. 143.
23 *The Kilmarnock Herald and North Ayrshire Gazette* 9 March 1917 p. 3.
24 Ayrshire Archives BA/18/53 (I am grateful to Katy Brown, Clerical Assistant (Archives) for her searches).
25 *The Kilmarnock Herald and North Ayrshire Gazette* 16 March 1917 p. 3.
26 *The Kilmarnock Herald and North Ayrshire Gazette* 16 March 1917 p. 3.
27 *The Kilmarnock Herald and North Ayrshire Gazette* 30 March 1917 p. 3.
28 Glasgow City Library archives SR22/74/5 County Chief Constables Office Report Book No. 10 20 April 1917 p. 123.
29 *The Scotsman* 31 March 1917 p. 7.
30 Glasgow City Library archives SR22/74/5 County Chief Constables Office Report Book No. 10. 11 January 1918 p. 138.
31 Robson, 1984, op. cit. p. 71.
32 Smale, D. 2018. The First World War and Policing in the Scottish Borders. *History Scotland* 18(1).
33 *The Berwickshire News* 16 January 1917 p. 7.
34 Smale, 2018, op. cit. p.34.
35 *The Southern Reporter* 23 January 1919 p. 5.
36 *The Jedburgh Gazette* January 31, 1919, p. 4.

37 Smale, 2018, op. cit. p. 34.
38 Emsley, C. 1996. *The English Police: A Political and Social History*. 2nd Edition. London: Longman.
39 Clapson, M. and Emsley, C. 2011. Street, Beat and Respectability: The Culture and Self-Image of the Late Victorian and Edwardian Urban Policeman. In *Police and Policing in the Twentieth Century*, ed. C. A. Williams, Chapter 14. Farnham: Ashgate Publishing, pp. 293–317.
40 Glasgow City Police Records SR22/55/17-24. Available from The Mitchell Library Archives, Glasgow.
41 Smale, D. 2007. The Development of Policing in the Scottish Borders c.1840–1890. Unpublished PhD Thesis, The Open University.
42 *The Berwickshire News* 20 March 1917 p. 3.
43 *The Stonehaven Journal* October 26, 1916, p. 6.
44 *Southern Reporter* 15 June 1916 p. 6.
45 HC Deb 22 February 1917 vol 90 cc1510 and 1543–4.
46 *The Economic Journal* 1916 26(101) pp. 110–2.
47 *The Berwickshire News* January 30 1917 p. 2.
48 *The Berwick Advertiser* January 12 1917 p. 3.
49 *The Berwickshire News* February 6 1917 p. 7.
50 *Daily Mail* February 22 1917 p. 2.
51 HC Deb 23 February 1917 vol 90 cc1606.
52 *The Berwick Advertiser* 2 March 1917 p. 8.
53 *The Jedburgh Gazette* 9 March 1917 p. 3.
54 *The Police Review and Parade Gossip* 16 March 1917 p. 81.
55 *The Berwick Advertiser* 16 March 1917 p. 3.
56 *The Jedburgh Gazette* 23 March 1917 p. 3.
57 *The Berwick Advertiser* 16 February 1917 p. 5.
58 *The Police Review and Parade Gossip* 30 March 1917 p. 104.
59 *The Hawick Express and Advertiser* March 16 p. 3.
60 *The Southern Reporter* 19 July 1917 p. 7.
61 *The Southern Reporter* 8 August 1918 p. 3.
62 https://en.wikipedia.org/wiki/Coatbridge (accessed 14/1/2024).
63 *The Coatbridge Leader* 18 January 1919 p. 3.
64 *The Coatbridge Leader* 21 April 1917 p. 3.
65 https://www.scottishpolicemedals.co.uk/the-king-s-police-medal-recipients-in-scotland/the-king-s-police-medal-for-distinguished-conduct/#:~:text=John%20Carmichael%2C%20the%20son%20of%20a%20policeman%2C%20was,in%201892%20and%20Deputy%20Chief%20Constable%20in%201894 (accessed 14/1/2024).
66 *Dundee Evening Telegraph and Post* 13 April 1917 p. 3.
67 *Dundee Courier* 19 March 1917 p. 1.
68 *Dundee Evening Telegraph and Post* 12 March 1917 p. 3.

69 HC Deb 15 March 1917 vol 91 cc1258–9.
70 *Dundee Evening Telegraph and Post* 12 March 1917 p. 3.
71 https://en.wikipedia.org/wiki/Edinburgh#Demography (accessed 14/1/2024).
72 *The Scotsman* 1 November 1919 p. 7.
73 *The Scotsman* 2 March 1917 p. 7.
74 *The Scotsman* 3 March 1917 p. 7.
75 *The Scotsman* 15 March 1917 p. 7.
76 Edinburgh National Archives SL260/ECP/1/2/1/10, 5 May 1917 General Orders No. 853.
77 *The Police Review and Parade Gossip* 12 April 1918 p. 118.
78 *The Police Review and Parade Gossip* 13 July 1917 p. 222.
79 Edinburgh National Archives SL260/ECP/1/2/2/5.
80 *The Scotsman* 21 April 1917 p. 9.
81 *The Scotsman* 16 March 1917 p. 3.
82 https://el4.org.uk/parish/haddington/ (accessed 14/1/2024).
83 Edinburgh Archives E65 SL260/ELC/1/4/2.
84 https://www.electricscotland.com/history/forfar/chapter26.htm (accessed 14/1/2024).
85 Angus County Council Standing Joint Committee minute book ACC3/2/5.
86 https://en.wikipedia.org/wiki/Greenock (accessed 14/1/2024).
87 https://www.greenocktelegraph.co.uk/opinion/14019776.police-experiment-paid-off/ (accessed 14/1/2024).
88 *The Police Review and Parade Gossip* 23 March 1917 p. 91.
89 *The Police Review and Parade Gossip* 5 April 1917 p. 106.
90 https://en.wikisource.org/wiki/1911_Encyclop%C3%A6dia_Britannica/Kincardineshire (accessed 14/1/2024).
91 *The Stonehaven Journal* 22 March 1917 p. 2.
92 Aberdeen City and Aberdeenshire Archives ASKcty/2/1/6 p. 125 (I'm grateful to Ashleigh Black, Archive Assistant for locating the files).
93 County of Kincardineshire Standing Joint Committee Minute Book No. 2. ASKcty/2/1/6 p. 135.
94 https://en.wikipedia.org/wiki/Montrose,_Angus#Demography (accessed 14/1/2024).
95 https://en.wikisource.org/wiki/1911_Encyclop%C3%A6dia_Britannica/Montrose (accessed 14/1/2024).
96 *Montrose Standard and Angus and Mearns Register* 12 January 1917 p. 5.
97 *Montrose Standard and Angus and Mearns Register* 9 March 1917 p. 5.
98 *Montrose Standard and Angus and Mearns Register* 16 March 1917 p. 7.
99 *Montrose Standard and Angus and Mearns Register* 30 March 1917 pp. 4–5: 27 April 1917 p. 5.

100 Dewey, P. E. 2014. *British Agriculture in the First World War*, 2nd Edition. London: Routledge, pp. 49–50.
101 *The Stonehaven Journal* January 25 1917 p. 2.
102 HC Deb 22 February 1917 vol 90 cc 1543.

CHAPTER 8

Release of Policemen in Birmingham and Glasgow: 2 Case Studies

Birmingham and Glasgow were outliers in the release of police as ploughmen. The former due to the length of the release, the tussle to recruit special constables and the city's arrangements for release of policemen into agriculture shown as a national example in the police weekly journal. The latter was targeted by Neville Chamberlain in his first visits to cities around Britain to recruit National Service volunteers and was subsequently hailed as an example for other local authorities and businesses. In Birmingham, release was announced on 20th March 1917:

> POLICE ON THE LAND. A number of Birmingham policemen are to be temporarily released from duty for urgent work upon the land.
>
> The Chief Constable (Mr. C. H. Rafter) has issued an appeal to all members of the force asking men who are skilled in agricultural work to volunteer to return to the land, and the response has exceeded the anticipations of the authorities. [1]

This announcement in the local press in mid-March showed the Chief Constable asking for only men skilled in agriculture to volunteer. Around 100 policemen were released full-time and 83 were still on the land on 3rd October, with agreement of the Watch Committee they remained until at least the end of July 1918 [2, 3] and were praised for their work by farmers [4]. No other force was found which lent policemen to farmers continuously for this length of time. Birmingham was also the home of

M. Fraser, *The British Police and Home Food Production in the Great War*, https://doi.org/10.1007/978-3-031-58743-6_8

Neville Chamberlain, recruited by Lloyd George as Director of National Service in December 1916 from his post as Lord Mayor.

Glasgow, the other outlier, was held as a leader following Neville Chamberlain's visit to the city on 27th and 28th February to plead for substitutes, particularly into agriculture in the name of patriotism. Harling Turner praised Glasgow's release of policemen, published nationwide[1]:

> THE VICTORY FURROW. The splendid lead given by Glasgow Corporation in releasing men in their employment possessing skill in the art of ploughing to assist farmers in the next few weeks has been followed by Edinburgh Corporation and by the Chief Constable of Roxburgh, Selkirk and Berwick.
>
> Glasgow has released over 80 of its municipal employees for this urgent work, and another squad of about 150 will be available within a few days. [5]

This provides an example of how other major locations, both urban and rural had responded. By giving the numbers of employees released, it provided a benchmark for others to emulate. Each location will now be developed in turn to show their unique contribution.

Birmingham. The Chief Constable from 1899 to 1935 was Charles Haughton Rafter, described by Klein as a baron, military in experience and style. He was a pioneer in educating his men but used individual's reluctance to improve their level of education as a threat to promotion prospects and showed frustration at non-participation. His autocratic style was seen particularly in the 1918 police strike, when he refused to admit that his men had serious grievances and remained remote from them [6]. However, he fiercely defended the men against recruitment into the military during the First World War, as the following paragraphs demonstrate. A number of factors occurred simultaneously leading to the release of police as ploughmen: recruitment of special constables to release regular policemen into the army, threatened strikes by agricultural

[1] The other newspapers to publish a similar report were: *The Midland Telegraph* 14 March 1917 p. 2; *Daily Gazette for Middlesbrough* 15 March 1917 p. 3; *Aberdeen Press and Journal* 14 March 1917 p. 3; *Leeds Mercury* 15 March 1917 p. 3; *North Wilts Herald* 16 March 1917 p. 7; *Lincolnshire Echo* 16 March 1917 p. 2; *Kilkenny Moderator* 17 March 1917 p. 6; *Cheltenham Chronicle* 17 March 1917 p. 3; *Kington Times* 17 March 1917 p. 3; *Cheltenham Looker-On* 17 March 1917 pp. 5–6; *The Southern Reporter* 22 March 1917 p. 7; *The Birmingham Daily Post* 19 December 1916 p. 2.

labourers and fear by farmers of being unable to comply with Board of Agriculture demands due to labour shortages, and the potato famine; this combination influenced the release of policemen into agriculture.

On 6th March 1917 the authorised strength of the force was 1,431 with 399 vacancies (27.9%) [7]. But Rafter had been publicly criticised in *The Birmingham Daily Mail* since November 1916 for not releasing policemen into the army, at a time when few men wanted to go to war and scarcely any employers wanted to release eligible men:

> The war-policy of the city, as applied to the police, is more than unsatisfactory. Even with a depleted force, we still have the scandal of a large number of men eligible for general service in the Army being retained – men who on all grounds ought to go to the front, and could go were there the effective will to send them. [8]

As Chief Constables retained responsibility for deciding the release of policemen into the army, Rafter was severely criticised in the local press for not supporting the calls for fit men of military age. Criticisms extended to the lack of recruitment of special constables and disallowing the few who had been secured to work on the beat; newspapers estimated only around 100 special constables were in place [9]. A campaign by the local press for more than three months to entice the Lord Mayor to hold public meetings to recruit special constables was followed by a report by Councillor Thomas Brown to the Watch Committee on 6th March 1917:

> the need for further special constables has become more urgent. The Lord Mayor recently called a public meeting at the Town Hall and made a strong appeal to eligible citizens to enrol for this service. The result has been eminently satisfactory. Mr. J. Ernest Hill (Prosecuting Solicitor), who has been appointed Head Special Constable, reports that he has received offers for service from over 3,500, and that 2,500 have already been sworn in. Mr. Hill is of opinion that there will be no difficulty in obtaining .... at least 5,000 special constables. [10]

Ernest Hill, the prosecuting solicitor to Birmingham police, was trusted by the local authority and respected by potential special constables; by the end of April 4,000 had been recruited [11]. This ended condemnation of the Chief Constable extending over the previous three months for being ineffective at recruiting special constables so that more policemen of military age could be released into the army, blaming

Rafter for having no willingness to release his men and comparing Birmingham unfavourably with other major English cities. Letters to the press by current special constables speedily followed, confirming they felt unwanted and that policemen of military age were hiding from military service under the blue uniform.[2]

The army were also aware of policemen of military age remaining in the force. On 4th December 1916 Rafter showed the Judicial Sub-Committee of the Watch Committee a communication he received from Southern Command asking for more policemen. The sub-committee recommended he send this to the War Office, to apply to all police forces, indicating Birmingham's willingness to comply [12].

On the appointment of Alderman A.D. Brooks as Lord Mayor following Chamberlain's resignation, recruited as Director of National Service in December 1916, Brooks published an open letter appealing for special constables, saying the Watch Committee would release more constables into the army but was unable without an increase in special constables [13]. It appeared 2 days after an editorial in *The Birmingham Daily Post* on 2 January (p. 4) criticising Rafter for writing privately to the volunteer force, around 1,500 men guarding munitions works, asking them to undertake police beat duty, which the editorial deplored, pressing for the open enrolment of special constables. Two further letters highly critical of the Chief Constable's lack of seriousness in recruiting special constables ensued, implying he was resisting the release of eligible policemen and the force's disrespectful treatment of existing special constables [14]. Further accusations of secrecy about the number of policemen of military age were also published "why, even now, should such great care be taken not to divulge the actual number of eligible policemen in the force?" [15].

In the face of sustained criticisms, the Lord Mayor held a public meeting in the Town Hall on 9th January to recruit special constables. *The Birmingham Daily Post* urged everyone to rally behind the Lord Mayor, estimating the numbers needed were around 3,000 to allow between 9 and 10 for each regular constable, estimating that between 500 and 600 policemen of military age still remained in Birmingham's police:

[2] For example, letters appeared in *The Birmingham Daily Post* 27 November 1916 p. 2; *The Birmingham Daily Mail* 27 November 1916 p. 4; *The Birmingham Daily Post* 28 November 1916 p. 2; *The Birmingham Daily Post* 29 November 1916 p. 2; *The Birmingham Daily Post* 5 December 1916 p. 7.

> The public have called for the release of eligible policemen. Inferentially they have accepted a great responsibility, which must now be fulfilled if the city is to recover its proud reputation for public spirit and patriotism. [16]

The appeal for a large attendance was to reduce adverse comments on Birmingham's lack of patriotism by the police and the disgrace this had brought on the city. The special constables should also be organised as a separate body, not as part of the regular force, and be supplied with appropriate clothing and a boot allowance [17].

The *Birmingham Daily Post* reported Alderman Brooks proposed scheme was "in every way satisfactory," to relieve the regular police of much patrol work; they would receive suitable clothing, a boot allowance and be compensated for any injuries sustained while undertaking the work. The meeting chaired by the Lord Mayor supported by the Chief Constable required attendees to complete the distributed enrolment form [18]. *The Birmingham Gazette* added that four or more hours equivalent to or in excess of 2 periods of duty per week was needed, and every effort would be made to place volunteers near to their home to work the beats of constables released for army service [19].

Police Orders show the Chief Constable moved speedily to request superintendents check the completed forms complied with military exemption regulations and applicants' "character is satisfactory" and they were not of "enemy extraction" [20], responding to anti-German antagonism. By 22 January 1,500 had enrolled, they would start patrol work in February [21]. On 5th February around 2,000 were to be sworn in in batches of 50 to begin work immediately afterwards [22, 23]. By 28th February 3,500 special constables had been recruited [24], ending the long-running and damaging dispute about enrolment of special constables and their allocation to routine police duties on the beat.

The compensatory release of military age policemen was agreed at the Special Judicial Sub-Committee of the Watch Committee on 26 January, attended by Colonel Sturgess and Lieutenant Ryland, who expressed the hope that when the special constabulary was at full strength, a considerable number of regular policemen would be released, assuring the committee there was no suggestion the police were unnecessarily retaining men of military age. Agreement was reached that constables would be released shortly, and that Birmingham would be a leader by writing to the Home Office requesting an order be issued to all police forces to release likewise [25]. Newspapers assured readers of agreement between the Lord Mayor and the Chief Constable that around 50 single policemen would be released into the army shortly [26, 27]. On 31st January *Police Orders* show the Watch Committee agreed 50 men offered to the military for mounted duty. A further 50 single men under 25 years of age who joined the police since the beginning of the war, supplemented by older single men would be identified by sending them for medical examination. On 9th February Rafter sent the lists of eligible men to superintendents requesting they send into the army all who had joined the police since the beginning of the war, responding to national claims of recruits enrolling in the police to avoid war service, and to choose from the remainder those to be conscripted. Superintendents were also asked to identify who was indispensable and the reasons—their cases would be made to a tribunal [28]. By the beginning of February newspaper reports show young single policemen from A to E divisions released into the army, but criticised some young men allowed to join the special constabulary, suspected of avoiding army service [29]. *Police Orders* give the call up date as 1st March, the special constables allocated to each division started work on 3rd March. By May 1917 release into the army was by quotas issued by the Home Office, resulting in a further call up of 28 policemen of medical category A[3] in early May [30].

[3] British Army Medical categories in 1914 were:

A. Able to march, see to shoot, hear well and stand active service conditions.

A1 Fit for dispatching overseas, as regards physical and mental health, and training
A2 As Al, except for training
A3 Returned Expeditionary Force men, ready except for physical condition
A4 Men under 19 who would be Al or A2 when aged 19

B. Free from serious organic diseases, able to stand service on lines of communication in France, or in garrisons in the tropics.

B1 Able to march 5 miles, see to shoot with glasses, and hear well
B2 Able to walk 5 miles, see and hear sufficiently for ordinary purposes
B3 Only suitable for sedentary work

C. Free from serious organic diseases, able to stand service in garrisons at home.

C1 Able to march 5 miles, see to shoot with glasses, and hear well
C2 Able to walk 5 miles, see and hear sufficiently for ordinary purposes
C3 Only suitable for sedentary work

D. Unfit but could be fit within 6 months.

Concurrently, members of the agricultural labourers' union threatened to strike if more farm workers were substituted by C3 men, farmers also strongly protested at the War Office national proposal to remove a further 30,000 farm workers into the army (see Chapter 3, p. 51). Birmingham's War Agricultural Committee (WAC), chaired by C. Napier Clavering, JP, expressed consternation, sending a resolution to Roland Prothero, President of the Board of Agriculture:

> Farmers in this district are staggered by the proposal to take more men for the Army, making it impossible to maintain cultivation. C3 men are useless as substitutes. The absolute collapse of the industry is feared.

They insisted that even if farmers trained C3 men, they would not become proficient for 12 months, and currently, with implementation of the Bath formula,[4] even four would not equal one experienced farm worker [31, 32]. However, from mid-April a deputation of WAECs meeting with Lloyd George agreed an increase in farm labour by women and soldiers would not be counted in the number of farm workers to be retained on farms according to the Bath formula [33]. By 22nd January in front page news, *The Birmingham Gazette* told its readers the call up of exempted farm workers had been stopped (p. 1). But protests continued from the Central and Associated Chambers of Agriculture meeting in London which passed a resolution that agricultural labourers should not

D1 Regular RA,RE, infantry in Command Depots
D2 Regular RA,RE, infantry in Regimental Depots
D3 Men in any depot or unit awaiting treatment

[4] The Bath formula:
"The following scale has been agreed between the Army Council and the Board of Agriculture and Fisheries *as a general guide* for determining the number of men to be retained on farms, namely:-
One skilled able-bodied man or lad (wherever possible not of military age) for each of the following:-
each team of horses required to cultivate the land;
every 20 cows in milk, when the assistance of women or boys is available;
every 50 head of stall or yard stock, when auxiliary feeding stock is resorted to and the assistance of women or boys is available;
every 200 sheep, exclusive of lambs, grazed on enclosed lands;
every 800 sheep running on mountain or hill pasture."

have been called up before adequate arrangements for their replacements were in place [34].

The strength of feeling is seen by fierce local protests at the further reduction of farm labour, an example was reported from Warwickshire Standing Joint Committee, when a prominent agriculturalist told the Chief Constable he would rather lose a policeman into the army than a shepherd [35]. The work of the police was trivial—they were only visiting cinemas—which could be undertaken by special constables or retired people [36].

Substitution of agricultural workers nationally by volunteers or members of local authorities was indicated to War Agricultural Committees on 27th January 1917 in a letter from Prothero as part of their reconstitution as Executive Committees (WAECs) [37]. The creation of WAECs with extensive powers of compulsion caused unease amongst farmers due to tenancy agreements which prevented ploughing grassland; insufficient labour; potential loss of income from change of land use; and compulsion to plough land likely to give a poor yield. But provided WAECs appointed competent members and maintained good links with the Board of Agriculture and the Food Production Department, farmers would have confidence in their work [38]. Composition of Birmingham City WAC was published on 10th February, with C. Napier Clavering as chair and Ruth Crosskey appointed to organise women's farm labour [39]. Prothero's visit to Birmingham also in February to discuss agriculture with midlands agriculturalists, saw farmer's main concern was lack of labour with worry that women workers and National Service volunteers would not be available for spring cultivation. Farmers would do their best but needed very considerable reinforcements of labour to avoid the loss of crops and the increases in fallow of the previous year [40].

Further emphasis on the urgent need for home grown crops came when the potato famine hit Birmingham by mid-February with predictions that none would be available in the next few days [41], evident from 23rd February [42]. From early March the few that were available had to be hunted down and when found retailers limited each shopper to 2lbs [43]. Representatives from Birmingham's Wholesale Market attended a conference in London to press for release of potatoes from growers, while in Birkenhead and Cardiff reports were of the police being deployed to maintain order in queues for potatoes [44, 45]. Questions in the House of Commons and an editorial in *The Birmingham Daily Post* confirmed the potato famine would continue until late spring or early summer and pleaded with everyone to eat any supplies sparingly [46], prompting appeals by both Lloyd George and Arthur Lee[5] to farmers to overcome all obstacles to produce the maximum crop of wheat, oats and potatoes in the next few weeks [47]. The lack of potatoes resulted in huge demands for turnips and swedes, with a consequent large price increase [48] and by 26 March reports told of hundreds of women raiding greengrocers' shops in Atherstone, 20 miles north-east of Birmingham, for the small supply of potatoes [49].

As food supplies dwindled, the Lord Mayor wrote a passionate letter to all citizens "the Lord Mayor's appeal" to economise on food and prevent wastage, particularly bread and flour portrayed as due to world-wide shortages caused by poor wheat harvests, cargo ships being used for war work and German submarines sinking many remaining cargo ships containing wheat imports. Afternoon tea should be abandoned, and the Food Campaign Committee arranged demonstrations around the city to show the best substitute foods and how to cook them; a weekly menu would be published [50]. By the beginning of May public paranoia increased with prominent headlines on food supplies which accused government of not releasing the truth about the number of ships lost "SHIPPING LOSSES INCREASING: MORE ADMISSIONS BUT NO FIGURES." and "HALF TRUTHS FOR THE PUBLIC. NO FACTS ON LOST TONNAGE. SECRET SESSION ON SUBMARINES" [51].

Food shortages escalated, developing into labour unrest in Birmingham in February 1918 alerting the Chief Constable to ask superintendents to assess how many regular policemen and special constables would be available for duty in case of the need to quell a disturbance [52].

Help for farmers to increase arable production urgently began in Birmingham with the launch of National Service to provide substitutes into key industries, identifying agriculture as a priority (see Chapter 3, pp. 65). Widely publicised in Birmingham, particular note was made of instructions to army recruiting officers that government departments were to be combed out of men below 22 years of age, specifying policemen, firemen, teachers, post office workers and civil service staff [53]; combing out excepted men employed in vital occupations such as

[5] Director General of Food Production.

agriculture [54]. By 7th February National Service volunteers were particularly needed for agriculture and other crucial industries[6] and urgent appeals for ploughmen were prominent in the press from mid-February, identifying policemen as a skill source [56]. However, by 26th February recruitment by National Service into agriculture from other groups had produced no volunteers, despite publicity campaigns [57]. Birmingham's National Service Committee agreed to work with the WAC to make a strong appeal to all migrants from rural areas into the city to immediately offer their services as ploughmen [58], swiftly followed by the Lord Mayor's appeal to Birmingham Watch Committee to respond to the Prime Minister 's call for ploughmen. The committee agreed to ask superintendents to assess how many constables could undertake this work and to ask for volunteers, applicants would be given leave of absence in this work of national importance [59], the response was published in the *Birmingham Daily Post:*

> The Chief Constable of Birmingham has received a request from the government to liberate policemen for ploughing. ... and Mr Rafter's request for volunteers has received a satisfactory response. Every police division of the city has yielded a quota of men, and it is understood that within the next week or so they will leave Birmingham for temporary work on the land. There are many skilled agriculturalists in the city force, and a few hold medals as winners of county ploughing competitions. [60]

The newspaper made known the success in eliciting many agricultural volunteers in the city police who were ex-ploughmen, some praised as possessing high levels of skill.

Every police district submitted numbers for temporary work on the land; they would leave for agricultural work within the next week, every volunteer was chosen [61]. *The Police Review and Parade Gossip* said the Chief Constable received the request from the Home Secretary, showing its importance and Rafter's high standing nationally [62]. On 2nd April the Judicial Sub-Committee of the Watch Committee agreed to release 98 policemen until the end of the season [63, 64], the local authority

[6] Apart from agriculture, the industries identified were: aeroplane construction, building construction, coal mining, coke workers, docks, engineering, explosives factories, flour millers, general labourers, iron and steel workers, motor mechanics, munition works, national steel factories, national filling factories, ore mining, quarry workers, shipbuilding, transport services [55].

would continue to pay their police wages and be compensated by each receiving 25s per week if unable to return home at night and 15s if able to return home, payable under the National Service scheme to the Watch Committee. The volunteers, approved by the National Service commissioner for the area, would retain their police incapacity benefit if injured at work, and be provided with overalls. In addition, they would receive railway vouchers for travel to and from work. The urgency was such that the Chief Constable wrote to all divisional superintendents to ensure they were ready to leave at a moment's notice [65]. Subsequently a further allowance of 2s 6d would be paid to men with dependents [66]. All would work in the surrounding districts where possible. These generous arrangements were for the sowing season only, if further help was needed a separate application would be necessary [67]. Encouragement to continue with the release of policemen into agriculture came in amendments to the Military Service (Review of Exceptions) Bill on 2nd April 1917 when men engaged in agriculture since 31st March and approved by the Board of Agriculture, would not be called up for re-examination for military service [68], ensuring Birmingham's released policemen were safe from conscription. Birmingham's arrangements became the national standard, published in *The Police Review and Parade Gossip* as advice to other forces and the WAC applauded police arrangements, but farmers had only applied for a few by 14 April [69].

The small number of policemen arrived on the land mid-March, in addition to 300 soldiers with agricultural and steam plough experience deployed around Warwickshire [70] during WAEC reports that around 500 owners of steam ploughs were not using them fully, 7-days a week from dawn to dusk and were placing obstacles to their use. In a survey during the previous four days, not one acre of land of between 15,000 to 20,000 acres had been ploughed. The Food Production Department expressed concern and the WAEC threatened stronger measures if urgent improvement was not made [71]. Police on the land would ensure surveillance of WAEC instructions and farmers' actions.

The police in Birmingham gradually recaptured their reputation for public service as, in addition to police release for agricultural work, 100 were also released to work in munitions, while around 4,000 special constables were at work and the numbers continued to grow. The Lord Mayor was now able to inform the public that the number of policemen released into the army was 450 [72]. However, further combing out occurred in June:

Military Authorities (Confidential)
Recruiting officers from each Sub Area in B'ham (sic) will call in due course at the various Police Stations to interview the officers in charge. Every information should be afforded to them, which is in possession of the Police, with regard to any person who may be supposed to be qualified for Military Service and .. still in civil life in the Divisions. (highlighting and underlining supplied) [73]

There was no hiding place. Furthermore, in the German Spring Offensive on the Western Front (see Chapter 3, pp. 73) the Home Office requested a further quota of 150 were released into the army on or before 20th April 1918 [74]. However, those remaining in agriculture retained ineligibility for call up. By 31 July from an authorised strength of 1,431, Birmingham contained 858 officers with a vacancy rate of 573, a loss of 40% of the force, however, it's unclear whether the number released into agriculture are contained within these numbers of staff in post.

Birmingham continued with the release of 89 policemen into agriculture and also allowed a further unspecified number [75]. In October 1917 the Board of Agriculture requested the further release of policemen into agriculture; the 83 still on the land were given permission to remain on the same conditions as before [76, 77]. However, the Watch Committee no longer agreed farmers should pay the minimum for police help, from 31st October they were required to pay 35s. per week to retain them [78]. These arrangements allowed police of military age released into agriculture to escape public notice and criticisms of evading military enlistment, while providing a much-needed public service in a time of national crisis.

Glasgow. The Chief Constable of the City of Glasgow police from 1902 to 1922 was James V. Stevenson [79]. The establishment was one-third of the total number of police in Scotland at 1,996 [80]. By the end of 1915, 692 had joined the military and around 3,000 special constables undertook police duties guarding vulnerable points such as water and gas works, power stations, tramway depots and important bridges [81]. By the close of 1917 the force was understrength by 723; 36.2% of its establishment [82].

A precursor and alongside Glasgow's release of policemen into agriculture was the acute suffering and protests by the city's working classes from the end of February 1917 as a result of the potato shortage. The crisis formed many column inches in national and local newspapers, increasing public anxiety. The *Glasgow Citizen* told readers in a prominent column that the secretary of the local Potato Merchants Association had wired the War Cabinet, and the Food Controller said stocks of potatoes were exhausted [83]. Two days later an equally prominent article said the Secretary for Scotland and representatives of the Food Controller had agreed to meet a deputation from the Glasgow and the West of Scotland Potato Trade Association at the Scottish Office in London [84], showing the significance of the crisis. A combination of reports of shortages in Scotland, the deputation and price fixing by the Food Controller raised questions in the House of Commons [85]. Public awareness of the shortage continued into the following week when *The Evening Times* told its readers:

> The scarcity of potatoes in Glasgow is becoming daily more acute. The very small stocks which local traders have in hand are being steadily depleted, and particularly no fresh supplies are coming in ... The position ... is bound to be very serious next week. All who are interested in the trade are awaiting anxiously the proposed action of the government regarding which it is expected some announcement will be made shortly. [86]

A written answer to a further question in the House of Commons on 12 March said the potato question in Scotland had been referred to a committee and an announcement would be made shortly [87]. The House of Commons continued to debate the shortage, sourcing, price and distribution of potatoes across the UK throughout March.

But Glasgow's working-class women were not willing to wait any longer for local and national government to provide a solution. On 16 March the Glasgow Labour Party Housing Association was reported to have organised a deputation, mainly of women, to the Corporation [88]. However, although they waited outside the chamber on 22 March, the deputation was refused, voted down by 47 votes to 19 with reports of "lively scenes" at the Corporation meeting [89], to the fury of the women when they heard that the Provost had "airily and impudently suggested that the demonstrators should go home and look after their babies" [90], reported in *The Glasgow Herald* [91] and on the front page of the socialist *Forward* [92] which portrayed the situation as a class war, telling readers that the Provost was a grain merchant and ship owner and calling the situation a potato famine. Newspapers reported the "violent scene" inside the Chamber with Councillors Dollan, Shinwell, Denny and Charlton suspended and Baillie James Stewart giving the

Provost "a minority lecture" on his insult to the working class. Outside the Chamber the suspended Councillors addressed the large number of women in George Square; many also remained waiting to "give the Lord Provost a "reception", but he eluded them," [93]. Councillor Shinwell told the crowd that an attempt would be made at the next meeting to secure a hearing for the women's deputation. Two further demonstrations were held, one on Glasgow Green on 25 March at the further refusal of the Corporation to hear the deputation and to protest at the language used by officials about them. A second was planned for 29 March at which the women and their children would march from the Royal Exchange in Queen Street to the Council Chambers (0.1 miles) to demonstrate that their deputation be heard [94]. Both *The Glasgow Herald* and *Forward* told readers that the Provost said there was no need for a deputation, as he had been in London to impress on government the urgency of the situation, showing his elite credentials while ignoring the working-class immediate shortages. His lack of empathy with the lower classes food poverty made the inequality clear, he was able to buy any food for his family and did not have to rely on potatoes. This type of example was reviewed in Chapter 5 as part of the 1917 report on industrial unrest nationally. It included the socialist newspaper's accusations of the provost as a grain merchant and ship owner with the widespread suspicion that while most of the population feared starvation, he and a few similar others were amassing huge profits from shipping.[7]

Politically, Glasgow was targeted by Neville Chamberlain in his recruitment of National Service volunteers. Starting on 17th February 1917, the week following the launch of National Service, he travelled to Bristol, London, Glasgow, Sheffield and Cardiff [95]. Arriving in Glasgow on 25 February he stayed two nights with the Provost. He met a number of prominent citizens at lunch on 26th at which he said he had identified, through the National Register of August 1915,[8] "upwards of 12,000 men over military age who gave as their secondary occupation—that was the alternative to the occupation they were then engaged in—some form of agricultural work" who were currently engaged in city jobs, and he appealed for their release [96]. This was greeted with delight and irony

[7] https://archive.org/details/industrialunrest00chanrich/page/6/mode/2up?view=theater (accessed 14/1/2024).

[8] See Chapter 2, p. 28.

by the Scottish farming community whose journal reported "Mr Chamberlain was telling us on Monday evening that there are 12,000 men in Glasgow who can render invaluable service on the farm. Can quite believe it" [97].

In the evening his reported rousing public address in St. Andrew's Hall asked for help for the National Service campaign by substituting agricultural workers, said to be well supported and the only city from the 5 he visited where he asked for substitutes into agriculture. He appealed to patriotism by men of non-military age, recognising that most men of military age were needed more by the army than at home. He also announced the appointment of Joseph Harling Turner[9] as Head of the Agricultural Department, to cheers from the audience. Following his visit, a half page in *The Glasgow Herald* published his speech [98] and promoted the case for helping agriculture, with an endorsement by the Directors of the Scottish Chamber of Agriculture to "do their utmost to increase the food supply" at their meeting on 1 March, congratulating the Prime Minister on his courageous policy.

The Glasgow Herald printed the Lord Provost's urgent appeal to every man in Glasgow skilled in ploughing and with agricultural experience, of whom he said there were many, to "help put land under crops." Portraying himself as a leader, he suggested that the many men with such experience employed by the Corporation would be released for agricultural work during the following two months and reported many volunteers had come forward. He also announced the establishment of a special committee of the Corporation to encourage the population to enrol in National Service [99]. Harling Turner's manpower recruitment for agriculture was also published in a leading farming journal on 3 March, endorsing the appeal to ex-ploughmen currently employed in city jobs to return to the land [100, 101]. Chamberlain's visit galvanised Glasgow, despite rowdy protests at an anti-industrial conscription meeting, which needed around 12 policemen to barricade the doors [102].

Chamberlain's visit resulted in the Provost, Sir Thomas Dunlop,[10] a senior partner in his father's shipping company, Thomas Dunlop and

[9] See Harling Turner's profile in Chapter 7, pp. 160–161.

[10] Sir Thomas Dunlop was Provost of Glasgow from 1914 – 1917 https://clandunlop.com/royalty-%26-chiefs (accessed 14/1/2024).

Table 8.1 Previous experience of agriculture including work with horses in Glasgow City Police in February 1917 [106]

Types of previous agricultural experience	*Numbers in each type of previous work*	*Numbers recruited into the Army or Navy*	*Numbers left in the police service*
Farm servants	250	98	152
Ploughmen	33	18	15
Gardeners	26	12	14
Blacksmith	17	11	6
Shepherd	15	7	8
Farm labourers	13	8	5
Gamekeepers	12	4	8
Ghillies	6	5	1
Crofter	6	1	5
Farmer	5	3	2
Stableman	4	1	3
Groom	4	1	3
Others: include Cattlemen, Byremen, Royal Horse Artillery, Mechanics, Cartwrights, and Groundsmen	7	2	5
Totals	**398**	**171**	**227**

Sons, and a Director of the Royal Bank of Scotland, created a baronet in 1916 [103], to recommend to the Corporation of Glasgow that:

> every man in their service with agricultural skill (ploughmen &c.) be granted leave during the next two months, and that the Corporation pay to such employees the difference between their present wages, and the wages and allowances received for such work; [104]

It was also noted by a Glasgow Councillor that thousands of policemen around Britain possessed agricultural skills [105]. The Corporation agreed that all heads of service should identify suitable staff, recognising that agriculture was a popular source of recruitment into the police service, verified in Table 8.1

Table 8.1 shows the large number of farm servants, who were likely agriculturally multi-skilled, also the many ploughmen. Although 171 had been recruited into the armed services, more than half remained in the police force.

The Corporation's Special Committee on Agricultural Produce on 1 March resolved to ask all heads of department, including the Chief Constable, to identify appropriate staff, each must give his approval for their release and should communicate directly with National Service to arrange placements which ensured entitlement to the government subsistence allowance [107]. *The Evening Times* reported 90 policemen, around 7% of the force, were released and that "The first batch of Glasgow policemen, a party of 28, who have volunteered for agricultural services, leave the city today ... for farms in Perthshire, Ayrshire and Kirkcudbrightshire." all said to have been former farm servants, showing release to locations distant from Glasgow, such as Perthshire in the east [108].

Glasgow was held as a pioneer across Britain, notably in Coventry, Leeds Cambridge and Lincolnshire [109]. The release showed cross-party support; the Liberal newspaper *The Scotsman* described Glasgow's release of police to plough as "the most gratifying result" [110] and in Cheshire, the dominant conservative approach of the *Cheshire Observer* gives Harling Turner's reference to the release as an example to others [111].

Prominent reports on 14 March said the Corporation was giving "every encouragement ... to the movement for helping agriculture" and that they would make up the difference in wages paid by the farmer, the men would also be entitled to a board and lodging allowance at government rates of 17s. 6d. per week if they worked away from home [112]. The Corporation's leading role to release policemen was seen in the wide publicity it received in the local press, including the Provost portrayed as the good person by asking the population for their understanding if the Corporation's services were not as efficient as usual during the period of the release [113].

Glasgow's release of policemen shows the class wars with claims by the protesting working classes that the Provost did not understand the urgency of their situation as he and those of a similar class could afford to substitute other foods for potatoes, whereas their families relied on them. He was deeply resented as being a ship owner and grain merchant who the working classes claimed were profiteering from the war, identified in the June 1917 Commission on Industrial Unrest [114].

In conclusion, the lengthy campaign by Birmingham's local press, insisted that special constables were recruited to undertake routine police duties to allow the release of more policemen of military age into the army. The arrangements for their release were advised to the Home Office to stimulate release in other police forces. Simultaneously, Birmingham's release of policemen into agriculture helped to overcome the previous stigma of policemen of military age being retained in the police service and helped, as well as providing surveillance over farmers, to increase home food production. The Birmingham arrangements for release were advocated as an example to other forces. Policemen were permanently released into agriculture until the end of the war and were welcomed by farmers which allowed surveillance of instructions to plough and plant crops in the national interest.

Glasgow's release similarly helped to pacify the population that the police with appropriate skills were helping to plough and plant crops in the national emergency, and that they were being sent widely across lowland Scotland. Following Neville Chamberlain's visit to promote National Service volunteering, Glasgow became a pioneer, their example cited in many other locations. However, initial unrest by the working classes demanded swifter action to alleviate their immediate suffering when the supply of potatoes failed and they experienced lack of sympathy for their plight by the majority of the corporation, demonstrating class wars in action. Their plight illustrates the findings of the 1917 inquiry into industrial unrest.

The differences between Birmingham, where the police were contacted by the Home Secretary to release policemen into agriculture, and Glasgow where National Service made the appeal are seen. This highlights the differences between the two systems identified in Chapter 3, p. 68.

The next and final chapter summarises why policemen were released into agriculture mainly in Spring 1917, and again for harvest and in Spring 1918. Their release demonstrated patriotism by helping farmers in the crisis of food insecurity, their skills in agriculture were welcomed by farmers, giving the police close access to a vital wartime industry in danger of collapse. Their immediate help showed police release into a key industry into which they were easily absorbed due to their prior skills, during a pending and actual national disaster. The next chapter also summarises the outcome from government control of agriculture in wartime, during a period when the population was demoralised and war weary with little progress despite[11] huge loss of life in battle and food shortages and inflationary food price rises at home.

References

1 *Birmingham Gazette* 20 March 1917 p. 3.
2 *Birmingham Mail* 3 October 1917 p. 4.
3 Birmingham Council Proceedings 1917–1918 L34.3. 31 July 1918 p. 583. West Midlands Police Archives, Birmingham.
4 *The Police Review and Parade Gossip*. 16 November 1917 p. 373 and 15 February 1918 p. 37.
5 *Daily Record* 15 March 1917 p. 2.
6 Klein, J. 2021. 'The best police officer in the force' Chief Constables and their men, 1900–39. In Stevenson, K. et. al. eds. *Leading the Police: A History of Chief Constables 1835–2017*. Abingdon: Routledge pp. 125–140.
7 Birmingham Council Proceedings 1916–1917 L34.3 p. 176. Birmingham City Archives.
8 *The Birmingham Daily Mail* 25 November 1916 p. 3.
9 *The Birmingham Daily Mail* 25 November 1916 p. 3.
10 Birmingham Council Proceedings 1916–1917 L34.3 p. 179.
11 *The Birmingham Daily Post* 28 April 1917 p. 6.
12 References by the Watch Committee and its sub-committees to the Chief Constable. Birmingham City Police Archives.
13 *Birmingham Gazette* 4 January 1917 p. 4.
14 *The Birmingham Daily Post* 4 January 1917 p. 3.
15 *The Birmingham Daily Post* 4 January 1917 p. 4.
16 *The Birmingham Daily Post* 9 January 1917 p. 4.
17 *The Birmingham Daily Post* 9 January 1917 p. 8.
18 *The Birmingham Daily Post* 10 January 1917 pp. 4 and 5.
19 *The Birmingham Gazette* 11 January 1917 p. 2.
20 *Police Orders* 8 July 1915 to 12 April 1917 10 January 1917. West Midlands Police Archives.
21 *The Birmingham Daily Post* 22 January 1917 p. 4.
22 *Birmingham Gazette* 30 January 1917 p. 6.
23 *The Birmingham Daily Post* 5 February 1917 p. 3.
24 *The Birmingham Daily Post* 28 February 1917 p. 8.

[11] I am most grateful to Corinne Brazier, Archivist, at West Midlands Police Museum for giving me access to their extensive police records.

25 References by the Watch Committee and its Sub-Committees to the Chief Constable. Special Judicial Sub. Com'e January 26th 1917, 8123–8126. West Midlands Police Archives, Birmingham.
26 *Birmingham Gazette* 2 February 1917 p. 6.
27 *The Birmingham Daily Post* 2 February 1917 p. 4.
28 *Police Orders* 8 July 1916 to 12 April 1917, 31st January 1917 and 9th February. West Midlands Police Archives, Birmingham.
29 *The Birmingham Daily Post* 3 February p. 4 and 10.
30 *Police Orders* 12 April 1917 to 12 June 1918. Birmingham Council Proceedings 1916–1917 L34.3, 3 July 1917 p. 407. West Midlands Police Archives, Birmingham.
31 *Birmingham Gazette* 19 January 1917 p. 5.
32 *The Birmingham Daily Post* 19 January 1917 p. 8.
33 *The Birmingham Daily Mail* 13 April 1917 p. 3.
34 *Birmingham Gazette* 24 January 1917 p. 5.
35 *The Birmingham Daily Post* 23 January p. 3.
36 *Birmingham Gazette* 24 January 1917 p. 5.
37 *Birmingham Gazette* 27 January 1917 p. 5.
38 *The Birmingham Daily Post* 27 January 1917 p. 6.
39 *The Birmingham Daily Post* 10 February 1917 p. 5.
40 *The Birmingham Daily Post* 16 February 1917 p. 4.
41 *Birmingham Gazette* 17 February 1917 p. 3.
42 *Birmingham Gazette* 23 February 1917 p. 3.
43 *Birmingham Gazette* 5 March 1917 p. 5.
44 *The Birmingham Daily Post* 5 March 1917 p. 8.
45 *Birmingham Gazette* 12 March 1917 p. 3.
46 *The Birmingham Daily Post* 6 March 1917 p. 4.
47 *The Birmingham Daily Post* 12 March 1917 p. 8.
48 *The Birmingham Daily Post* 14 March 1917 p. 8.
49 *The Birmingham Daily Post* 26 March p. 8.
50 *Birmingham Gazette* 26 April 1917 p. 2.
51 *Birmingham Gazette* 1 May 1917 p. 1.
52 *Police Orders* 12 April 1917 to 12 June 1918, 6th February 1918 p. 370 West Midlands Police Archives, Birmingham.
53 *Birmingham Gazette* 8 February p. 3 and 9 February 1917 p. 6.
54 *Birmingham Gazette* 6 February 1917 p. 5.
55 *The Birmingham Daily Post* Trades in need of volunteers. 7 February 1917 p. 8.
56 *The Birmingham Daily Post* 24 February 1917 p. 5.
57 *Birmingham Gazette* 26 February p. 2.
58 *Birmingham Gazette* 10 March 1917 p. 3.

59 Police Orders 8 July 1916 to 12 April 1917, 14 March 1917, West Midlands Police Archives, Birmingham.
60 *The Birmingham Daily Post* 22 March 1917 p. 8.
61 *Police Orders* 8 July 1916 to 12 April 1917, 23 March 1917, West Midlands Police Archives, Birmingham.
62 *The Police Review and Parade Gossip* 30 March 1917 p. 102.
63 References by the Watch Committee and its sub-committees to the Chief Constable para 8205.
64 *Birmingham Gazette* 5 April 1917 p. 3.
65 *Police Orders* 8 July 1916 to 12 April 1917, 4 April 1917, West Midlands Police Archives, Birmingham.
66 *Police Orders* 8 July 1916 to 12 April 1917, 9 April 1917, West Midlands Police Archives, Birmingham.
67 *The Police Review and Parade Gossip.* 13 April 1917 p. 119.
68 Military Service (Review of Exceptions) HMSO, 1917 30 March 1917.
69 *The Birmingham Daily Post* 14 April 1917 p. 4.
70 *Birmingham Gazette* 22 March 1917 p. 5.
71 *The Birmingham Daily Post* 22 March 1917 p. 8.
72 *The Birmingham Daily Post* 28 April 1917 p. 6.
73 *Police Orders* 12 April 1917 to 12 June 1918, 25 June 1917 West Midlands Police Archives, Birmingham.
74 *Police Orders* 12 April 1917 to 12 June 1918, 11 April 1918, West Midlands Police Archives, Birmingham.
75 Birmingham Council Proceedings 1917–1918 L34.3. 31 July 1918 p. 579 and 583. West Midlands Police Archives, Birmingham.
76 *Birmingham Mail* 3 October 1917 p. 4.
77 *The Police Review and Parade Gossip.* 2 November 1917 p. 350.
78 References by the Watch Committee and its sub-committees to the Chief Constable 9 October 1917 para 6608–9. West Midlands Police Archives, Birmingham.
79 Stallion, M. and Wall, D.S. (2011) *The British Police Forces and Chief Officers 1829–2012* 2nd edition Hook: Police History Society p. 107.
80 Report of the Committee on the Police Service of England and Wales (1919) Cmd. 253 p. 5.
81 Grant, D. 1973. *The Thin Blue Line: The Story of the City of Glasgow Police.* London: John Long.
82 City of Glasgow Police. Criminal Returns for the Year Ending 31 December 1917; p. 10 available from the city archives in the Mitchell Library Archives, Glasgow.
83 *Glasgow Citizen* 26 February 1917 p. 2.
84 *Glasgow Citizen* 28 February 1917 p. 2.
85 *HC Deb 06 March 1917 vol 91 cc 194–5.*

86 *The Evening Times Friday 9 March 1917 p. 2.*
87 *HC Deb 12 March 1917 vol 91 cc729–30W.*
88 *The Bulletin 16 March 1917 p. 2.*
89 *The Bulletin 30 March 1917 p. 6.*
90 *Forward 24 March 1917 p. 1.*
91 *The Glasgow Herald 16 March 1917 p. 8.*
92 *Forward 24 March 1917 p. 1.*
93 *Forward 24 March 1917 p. 1.*
94 *Forward 24 March 1917 p. 4.*
95 Grieves, K. 1988. *The politics of manpower, 1914–18.* Manchester University Press. p. 112.
96 *The Glasgow Herald* 27 February 1917 p. 8.
97 *The Scottish Farmer* 3 March 1917 p. 155.
98 *The Glasgow Herald* 27 February 1917 pp. 5 and 7–8.
99 *The Glasgow Herald* 2 March 1917 p. 8.
100 *The Glasgow Herald* 2 March 1917 p. 8.
101 *The Scottish Farmer* 3 March 1917 p. 155.
102 *The Sunday Post* 15 April 1917 p. 3.
103 The London Gazette. 1 September 1916. p. 8592.
104 Minutes of the Corporation of Glasgow November 1916 to April 1917. 1 March 1917 pp. 686–7, available from the Mitchell Library, Glasgow archives.
105 Police for Farm Work. Questions Discussed at Glasgow, *The Police Review and Parade Gossip,* 9 March 1917, 75.
106 *Glasgow City Police Records SR22/55/17, 18, 19, 20, 21, 22, 23, 24.* Available from the city archives at the Mitchell Library Archives, Glasgow.
107 *The Glasgow Herald* 2 March 1917 p. 8.
108 *The Evening Times* 16 March 1917 pp. 2 and 4.
109 see *The Midland Telegraph* 14 March 1917 p. 2; *Cambridge Daily News* 14 March 1917 p. 4; *Leeds Mercury* 15 March 1917 p. 3; *Lincolnshire Echo* 16 March 1917 p. 2.
110 *The Scotsman* 26 March 1917, p. 6.
111 *Cheshire Observer* 17 March 1917, p. 3.
112 *The Evening Times* 14 March 1917 p. 2.
113 *The Glasgow Herald* 2 March 1917 p. 5.
114 Chance, W. 1917. *Industrial Unrest: The Reports of the Commissioners (July 1917) Collated and Epitomised.* Published for the British Constitutional Association, London https://archive.org/details/industrialunrest00chanrich/page/6/mode/2up?view=theater (accessed 10/1/2024).

CHAPTER 9

The Outcome for Britain of the Food Shortages of 1917

This book has shown how the discourse of patriotism guided the state's efforts during wartime at home and on the battlefield. The British police, as representatives of the state, were expected to lead and support this aim, by giving an example to their local populations. Therefore, when Britain was blockaded and experienced food shortages, one of the major aims of the state at home was to feed the people from home-grown food. High ranking officials from the Prime Minister down to local mayors implored everyone to help in an urgent effort during Spring 1917. Pleading with everyone to help was published nationwide and could not be ignored. Each and every one should forego their traditional roles and focus on food production. Publicity of the needs for, and the population's worries, that they had insufficient food was made plain through increasing industrial unrest and particularly in the example of Glasgow where women demonstrated at the lack of a major staple item in their diets. The results of the police not following the state's aim was seen in Warwickshire, where the Chief Constable was told by a farmer that they should change their trivial activities, such as visiting cinemas, to focus on helping farmers to grow food.

Police support for patriotism is shown in the numerous examples in this book where they loaned manpower skilled in agriculture to local and, on occasions, to distant farms. Instances of patriotism to bolster local populations are seen everywhere, even in very small areas, such as Montrose,

M. Fraser, *The British Police and Home Food Production in the Great War*, https://doi.org/10.1007/978-3-031-58743-6_9

where public officials including the police, planted fields with potatoes in a joint effort. These examples were applauded, welcomed and publicised as a good news story locally and nationally to encourage others to help likewise.

However, an additional main aim of the state was to provide the army with a continuing number of recruits from fit young men, in which the police were a target due to their personnel profile. When they were seen to resist mobilisation, the press continually targeted them with protests, raising the profile of their lack of patriotism, as in Birmingham, until they capitulated and reorganised the force to contain large numbers of volunteers of older age; this occurred in every police force but most with less public pressure.

The second half of the war saw huge loss of life in battle, for Britain particularly on the Western Front, while at home fears of population starvation were realised when imports were curtailed, due to the removal of ships for the war effort and enemy action. Britain speedily focussed on crop production to try to provide for the Home Front and the fighting forces in France and Italy. However, government also needed to ensure military manpower and the struggle between government departments for all sources of labour increased as war progressed. Strategies to identify and subsequently retain and substitute manpower into key industries, showed government's increasing control over the workforce and is seen in Chapter 3. The use of all sources of labour, particularly from the army, to boost agricultural productivity resulted in limited success, at least initially, as army substitutes were often transient and lacked ability. A key measure by government in early 1917 to keep farmers onside was to guarantee minimum prices for crops to overcome decades of losses from arable farming.

Aims to help agriculture to produce sufficient food for the population also included innovative use of mechanical equipment, particularly petrol-driven tractors, to attempt to ease the scarcity of labour in farming. The role of the Food Production Department in supplying farmers with manpower, equipment and supplies was crucial and is developed in Chapter 4.

The effects on the mass of the population brought about by food shortages and frightening reports of food riots in enemy nations, alerted the people to threats of starvation and showed government and the police the real potential for food riots in Britain. Increasing industrial unrest was investigated in June 1917 with food insecurity found to be the major factor, an example of which is in Chapter 8 in Glasgow. The availability of staple foods and how government attempted to control their supply is included in Chapter 5.

Into this tense situation on the Home Front in early 1917, the year of war weariness, when the population became increasingly frightened at the potential for starvation, the control exerted on agriculture to produce an increasing supply of home grown crops saw the police released in many locations to help farmers to implement Board of Agriculture policies, delegated to county level in England and held by the Board of Agriculture in Scotland. Due to their agricultural experience they were easily absorbed into the farming workforce.

The last three chapters of this book shows how "Police as Ploughmen" and "Police for the Plough"[1] highlighted the release of constables into agriculture to help farmers to increase crop production, in a government-led attempt to feed the nation. The police throughout England and Scotland responded to this national crisis by providing competent additional manpower to supplement that already provided by the army, which was of dubious quality, threatening to destabilise efforts to increase home food production. The contribution of *The Police Review and Parade Gossip,* a widely read and influential weekly journal, flattered policemen who possessed the skills to plough by highlighting their abilities not only to help in a time of national crisis, but also by their backgrounds as elite agricultural workers. Ploughmen were amongst the highest paid farm workers for their skills in being able to hold the plough sufficiently in the ground to turn the soil, and also to control the horses at a time when working with animals was higher paid than other manual farm work. This high-profile encouragement by the journal included publication of the advantageous arrangements for release by other forces, for example in Birmingham:

> POLICE FOR FARM WORK. THE BIRMINGHAM ARRANGEMENTS. The question of the temporary release for service on the land during the present sowing season of members of the Birmingham Police Force who have been previously engaged in agriculture came up for consideration at a meeting of the Watch Committee ... leave of absence should be given to the Constables selected, and that any injury sustained whilst so

[1] The headlines in *The Police Review and Parade Gossip* appeared on March 23 1917 p. 91; March 30 1917 p. 102 and Answer 13,764 G. A. p. 97; and 5 April 1917 p. 106.

> employed should be regarded as having been sustained whilst on ordinary Police duty. ... constables should continue to be paid their existing rate of wages. The committee further recommended that a payment of 25s. per week should be paid to each of the men engaged continuously away from home, and 15s. each per week to men who were able to return to their homes at night. [1]

This guidance gave advice to other forces, it was designed to encourage similar volunteering across the journal's wide readership of policemen on the beat as well as Chief Constables, Watch and Standing Joint Committees of the types of arrangements they should consider. Inspiration was also fostered by reports in the local and national press of the numerous other locations where policemen had been released for ploughing.

The government-led urgency to help agriculture from late March 1917 saw a national campaign: the Home Secretary wrote to Chief Constables asking for their help, the Director of Food Production published half column advertisements in local and national newspapers entreating the release of ex-ploughmen from other occupations for the crucial six-week period to plough and plant crops for an autumn harvest, and local authorities released policemen, often at the behest of National Service, paid partly by the farmer while claiming a proportion of their wages from National Service.[2] These measures and others shown in this book (Chapters 6–8) demonstrate the serious and concerted effort by government to feed the nation from home-grown produce.

However, the police also benefitted from these arrangements. Policemen of military age substituted into agriculture were out of sight of local people as their visibility caused consternation when they were on beat duty on the streets, making them prone to verbal attacks for their lack of patriotism by not being in the army:

> The Constable told the Court that the woman accosted him with the remark. "Why aren't you in the Army? A young man like you ought to be

[2] It is difficult to know if these two sources completely covered the police wages from Spring 1917 to the end of the war. Each police force decided its own rates of pay for their constables and also the level of the police war bonus. Farmers similarly paid different rates for ploughmen. However, it appears in some of the Scottish Local Authorities that an excess was received which allowed a bonus to be paid to the released policemen and also to those whose workload increased as a result (see Chapter 7). The nearest estimate of police wages nationally can be seen in the 1919 Desborough Report, Appendix II [2].

> ashamed of yourself going about in a Policeman's uniform." He ordered her off but she replied that she was not afraid of him, and in the presence of a large crowd she repeated that he ought to be ashamed of himself, and used bad language. [3]

This quotation demonstrates to readers the likelihood of verbal assault on individual policemen with loss of public support for the service, if young fit men did not sign up for army service. During December 1916 the army called for further large numbers of recruits due to huge loss of life on the Somme and at Verdun, therefore tensions were high for all men of military age to respond to calls to patriotism by signing up, with threats by the army that if more men did not come forward the war would last for longer. It seems that the police were well recognised as a source of young fit men of military age, as shown in Chapter 1.

Further evidence of loss of public support for the police as a service by delaying recruitment into the army was seen in the public altercation with the press in Birmingham from November 1916 which lasted for 3 months (see Chapter 8). The police were highly criticised for being secretive, suspected of withholding the numbers of policemen of military age in their employment and by not recruiting special constables so that they could be released. Letters also appeared in the press in Scotland accusing the police of not releasing their fit young single men into the army, claiming they were indispensable while acting as shirkers [4]. In addition, the public judged that policemen of military age would provide a better service to the country by serving in the army than by remaining at home where crime was designated trivial and fighting it less important.

Therefore, for both individuals and the service, releasing policemen with appropriate skills into a protected industry such as agriculture was praiseworthy, restoring public confidence in patriotism and flattering individuals by recognising their skills. The release also brought the police closer to an industry in crisis and under pressure to feed the nation. It provided surveillance over errant farmers and disgruntled farm workers.

The benefits to individual policemen and to the service, until April 1918 with the Spring Offensive, were that policemen released into the protected occupation of agriculture would not be conscripted during the periodic comb-outs of the police. Therefore, they were less likely to sustain serious injury or be killed in war and would return to the service as experienced regular policemen at the end of the war, when it was expected the service would be reconstituted.

However, the police did not release most of their men into agriculture indiscriminately. Most were released on condition they were allocated to farms close to their police stations, so that they could be recalled if needed urgently, such as to control a riot, which the special constables were said to be unable to deal with. But many of those released from Edinburgh and Glasgow provided a service to farmers in distant locations throughout Scotland and Metropolitan policemen were also released into the counties adjacent to London.

This book shows the build-up to the national crisis of food shortages and threatened population starvation, with the ways government responded being pivotal. As Lloyd George said in his memoirs "the food question ultimately decided the issue of this war" [5]. Britain's response can be compared with enemy nations who paid less attention to feeding their home populations leading to serious rioting and social disturbance [6, 7]. The release of policemen into agriculture in Britain not only helped to plough the fields to prepare for spring planting, but their presence and acceptance by farmers would also help to ensure War Agricultural Executive Committee instructions were enforced, and manpower and equipment was deployed effectively, while minimising the risks of invasion by frightened townsfolk desperate to access supplies at source, as was said to occur in enemy nations.

Although other organised adult groups such as the Women's Land Army and Prisoners of War have received attention in published works, the contribution of the police to feeding the home population has not been made known until now. Indeed, their contribution occurred before any of these groups; they acted as first responders in the national emergency. Although the army imposed their unwanted members (formed into C3 Agricultural Companies) onto agriculture in early 1917 when the crisis reached its height, the police carefully selected those they released so that they could easily be incorporated into agriculture, so spreading policing out into the community in a time of great potential for civic unrest.

The numbers and proportions of each force released, where it is known, varied (see Chapters 6–8). This can be explained in some locations where there was unrest, as in Cheshire where farmers refused to plough or resisted directions by the War Agricultural Executive Committee

(WAEC), the police were released in larger proportions which helped farmers to carry out the WAECs plans while maintaining social order. The maintenance of order was assigned to the police in the County and Borough Police Act of 1856 in England and Wales, and in Scotland by the Police (Scotland) Act, both remained current during the First World War. Similarly in Roxburgh, Berwick and Selkirk where disruption occurred with farm labourers striking for increased wages, the police were released in the largest proportion seen, compared with other forces. In Scotland the Secretary of State agreed the release as approved service under the Police Act. However, this pattern did not occur everywhere, in Norfolk where farm labourers had been on strike periodically for the previous two years for increases in pay, the police were only encouraged to work in agriculture in off-duty time and their numbers are not known. In other locations no indications were found of disruption by the agricultural workforce during the period under study.

Periodic release was also requested by the WAECs during subsequent periods of increased workload: Autumn harvest and Spring ploughing and planting. In England it is not known how many forces released policemen for harvest 1917, although this was the first time the Metropolitan Police were released into agriculture, to the pleasant surprise of Board and the Prime Minister. But some forces refused, for example Colchester in Autumn 1917 and Chester in Spring 1918, but in Scotland the numbers doubled to around 500 in Autumn 1917 compared with those released in the Spring [8]. Following the Armistice in November 1918, imports of food did not immediately increase, as shipping was needed to demobilise troops and equipment [9]. However, no further release of policemen into agriculture was found after harvest 1918.

The agricultural returns for 1917 showed the modest success of the plough policy during its first year, with an increase of 238,000 acres of arable land growing 985 more tons of corn and 156 tons of potatoes [10]. The corresponding small decrease in permanent grass, around 250,000 acres had been ploughed, indicating that most of the increases in acreage devoted to crops were on previously disused or under-cultivated land or on fallow land or temporary pasture, much of which had been broken up between March and May 1917 [11], when policemen were first released.

Table 9.1 Land use in Britain 1914–1918 (in millions of acres)

	1909–13	*1914*	*1915*	*1916*	*1917*	*1918*
Tillage	10.46	10.38	10.38	10.23	10.57	12.36
Temporary grass	4.15	3.92	3.88	4.12	4.04	3.49
Permanent grass	17.45	17.61	17.58	17.49	17.25	15.90
Totals	32.06	31.91	31.84	31.84	31.86	31.75

From Dewey, P.E. (2014) *British Agriculture in the First World War*. 2nd edition. London: Routledge, p. 201

The increase in arable land as a result of the government-led Plough Policy compared with pre-war is demonstrated in Table 9.1 and the average increases per region are in Table 9.2.

Whereas Dewey shows in Table 9.1 the gradual decrease in tillage up until 1916 when government intervened to nationalise farming in the national interest, from 1917 he found a small increase by harvest of 187,420 acres of pasture ploughed in England and Wales. He asserts that the increase in arable land for 1918 compared with 1917 was considerable but estimates vary. Middleton gave the increase in England and Wales as 1,557,000 acres and in Scotland 227,000 acres, with a reduction in grassland of 2,219,000 acres in England, Wales and Scotland [12], while Ernle states in the UK as a whole approximately one million acres had been added to the ploughed area, which included large amounts of previously disused land [13], while Dewey said by the end of 1918, 1.86 million acres of tillage was added in England and Wales and 0.28 in Scotland compared with 1916 [14], despite a reduction of 94,000 acres of cultivated land in England and Wales and 15,000 in Scotland, some of which had been used for airfields and military camps, munitions and army supplies. Comparisons from 1880s, when land use was progressively given over to pasture with the wartime effort to increase arable is shown diagrammatically in Fig. 9.1.

Figure 9.1 shows the decline in tillage and compensatory increase in permanent grassland from the early 1880s until the plough policy in 1918 made an impact, which is discussed in Chapters 2 and 3.

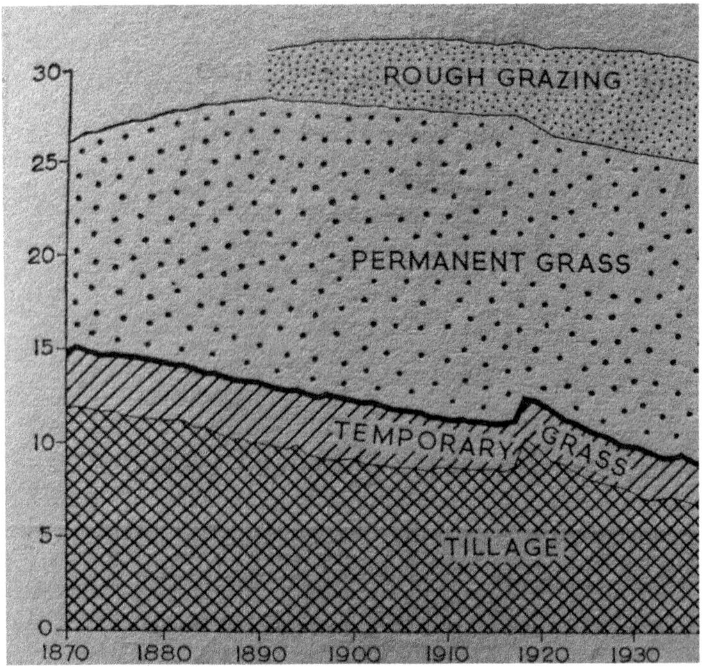

Fig. 9.1 Types of land use 1870–1935 in thousands of acres [15]

The varying contributions to tillage by region between 1916 and 1918 are given in Table 9.2.

Table 9.2 shows the smallest areas of increase in England are in the East where traditionally drier weather led to more emphasis on crop production, whereas the West increased its arable area during the war by larger amounts. These increases in acreage are the result of the widespread ploughing of permanent grassland in late 1917 in preparation for the 1918 harvest, compared with the 1917 programme, when mainly derelict and under-cultivated land and temporary pasture had been ploughed and brought back into cultivation. Yields showed no significant difference between crops sown in new or old ploughed land [16, 17].

The ability and willingness of farmers to comply with government directives was revealed in the limited amount of land dispossessed of owners or tenants. By the end of 1918, only 27,287 acres of badly farmed land had been taken into possession by the WAECs who arranged for

Table 9.2 Tillage increases in Britain 1916–18 (in acres)

*Region** *England*	*Acres*	*Increased tillage*
East	146,257	8.9%
North-East	166,904	9.3%
South-East	162,821	17.0%
East Midlands	183,990	23.1%
West Midlands	208,735	28.8%
South-West	264,165	37.2%
North-East	222,137	29.3%
North-West	242,421	39.9%
England total	1,597,430	20.0%
Wales	245,192	52.2%
Scotland	275,793	15.1%
British total	2,118,415	20.6%

*Board of Agriculture regions
From Dewey, (2014) op. cit. p. 203

cultivation, while 317 occupiers of a total of 20,197 acres were released from tenancies and the land placed under new management to increase productivity [18], demonstrating the persuasive powers of the WAECs to bring farmers on side.

By 1918 the government-assigned targets for each WAEC, despite the problems of labour and mechanical help for farmers, led to one of the largest amounts of grain planted between February to April. Agriculture ministers urged government to release more men from the army to harvest the gigantic crop of 7 ½ million acres of grain and potatoes in England and Wales [19, 20], Ernle estimated the addition was:

250,000 acres of wheat
75,000 acres of barley
850,000 acres of oats
32,000 acres of potatoes
5,000 acres of other root crops
14,000 acres of beans
15,000 acres of peas [21]

The weight of food produced for human consumption compared with 1914 was:

Table 9.3 United Kingdom net food production 1914–18 (billion calories)

	1900–13	*1914*	*1915*	*1916*	*1917*	*1918*
Cereals	4.24	4.63	5.36	4.40	4.80	6.83
Potatoes	3.15	3.46	3.54	2.36	4.32	4.37
Meat	6.13	5.83	5.85	5.67	4.90	3.78
Milk	3.02	3.08	2.83	2.62	2.31	2.09
Butter, cheese	1.14	1.19	1.19	1.17	1.08	0.99
Poultry, eggs	0.06	0.06	0.06	0.06	0.06	0.06
Cottage produce	2.65	2.65	2.65	2.65	2.65	2.65
Other vegetables	0.31	0.31	0.29	0.29	0.28	0.31
Fruit	0.18	0.20	0.18	0.17	0.20	0.14
TOTALS	21.06	21.41	21.95	19.39	20.60	21.22
As percentage	100	102	104	92	98	101

From Dewey, P.E. (1980) Food Production and Policy in the United Kingdom, 1914–1918. *Transactions of the Royal Historical Society* 80, 71–89.

14,017,000 tons in 1914
11,611,000 tons in 1916
18,007,000 tons in 1918 [22].

The huge increase in 1918 was despite cancellation of certificates of exemption from all farmworkers in April 1918 and the age for military service raised to 51 due to calls by the army to combat the German Spring Offensive [23, 24].

How increased production translated into calories to feed the population is given in Table 9.3. Despite the periodic food shortages, such as the lack of potatoes for around 3 months mid-1917, together with threats of population starvation promulgated in the press and food price inflation, the British population generally did not suffer from a greatly inferior quantity or quality of food, compared with before the war as Table 9.3 shows, despite periodic shortages.

Table 9.4 UK food supply 1909–18

	Average 1909–13	*1914*	*1915*	*1916*	*1917*	*1918*
Calories per average man per day	3442	3454	3551	3418	3320	3357
Ratio compared with 1909–13	100	100	103	99	96	97

From Dewey, (1988) Nutrition and living standards in wartime Britain. In Wall, R. and Winter, J. (eds) *The Upheaval of War: Family, Work and Welfare in Europe, 1914–1918*. Cambridge: Cambridge University Press p. 203

Table 9.3 gives a small increase in cereals in 1917 compared with 1916 with a considerable increase in 1918, while the potato famine in the winter and spring of 1916/17 turned into a glut by the end of 1917 due to the emphasis on growing potatoes. The change away from pastoral farming producing meat and dairy products, central to the plough policy, is seen in the decline in these products as war progressed. However, systems of food distribution and rationing were not implemented until early 1918, so that although the United Kingdom as a whole produced sufficient food, local shortages were evident, causing huge queues at shops and disturbances when supplies arrived, many requiring police control. But the average calorific intake remained remarkably stable as shown in Table 9.4.

The above Table demonstrates that the UK food supply remained almost unchanged during the war and is in stark contrast to other nations, particularly Germany and Austria-Hungary.

Ernle praised farmers' efforts, but expressed disappointment that they did not feel well supported by substitute soldiers or by equipment:

> During the first weeks of 1918 the weather was ideal for ploughing. Farmers seized their opportunity gallantly. They had received little of the promised help in men or machinery. But, working themselves in the fields, early and late, and strongly backed by the willing labour of the men and women they employed, they made a magnificent attempt to complete their task. Mainly from their own resources they accomplished more than two-thirds of the programme. [25]

We see the problems associated with substitution and supplies of equipment in Chapters 3 and 4. Lloyd George also praising the work of the food production campaign "without the extra millions of tons of home-grown food which it secured the nation would have gone hungry" [26].

From April 1917 the police acted as an organisation of social order in food production, carrying out the aims of the state in the discourse of patriotism. The mechanisms that maintained social order spread more directly out into the community in two ways: 1. By recruiting large numbers of special constables who undertook key positions in the community and 2. By involvement in the key industry of agriculture on the Home Front and making their presence known through local newspapers. By being well respected by farmers, many of whom they were likely to know through periodic delivery and collection of censuses and other war requirements, the police acted to pacify the terrified population, a master stroke of constructing social order that, provided sufficient food was produced, they showed the nation that starvation was unlikely [27]. In this sense, potential starvation is the enemy of social order. However, in the first year of help to farmers, as is seen in Table 9.3, very little difference was made to crop production and distribution remained uneven, but media reports that agriculture was a top priority and farmers were being helped in every way possible to produce more food, mainly pacified the population, although there were a few local disturbances and industrial unrest increased as is shown in Chapters 5 and 8.

This use of police as ploughmen shows the process of state power working through farming to pacify the population sufficiently to encourage continual production of the war effort at home. Through surveillance of food production by working alongside farmers, the police are seen as an institution of the exercise of administration of the population, rather than of the exercise of law.

Police as Ploughmen also bears modern comparisons. Just over 100 years after the agricultural crisis of early 1917, infections by the COVID-19 virus were said to threaten to overwhelm the NHS. January 2021 provides another example of the release of policemen into a service under intense pressure.[3] This was the period of maximum infections from the pandemic, and the London Ambulance Service was threatened to be overwhelmed not only by the number of emergency calls, but also by the many ambulance drivers who were either self-isolating, according to

[3] I'm grateful to Dr Rick Muir, Director of the Police Foundation for pointing this out.

government instructions, or were ill from the virus. Reports of an increase of nearly 50% in emergency calls, from 5,500 on a "typical busy day" to around 8,000, resulted in 75 Metropolitan police officers being transferred to drive ambulances and a further 125 placed on standby, available if the situation deteriorated further. The released officers were chosen for their qualifications to drive larger "blue light" vehicles, although one day's training was said to be necessary before they were deployed. The Mayor of London announced the situation in the London Ambulance Service as "a crisis" and declared a major incident: "If we do not take immediate action now, our NHS could be overwhelmed and more people will die" [28]. This example of the temporary release of police into a front-line service in order to maintain it has three major continuities with the 1917/18 crisis in food production:

1. The definition by the Mayor of a crisis situation;
2. the choice of policemen appropriately qualified for the work;
3. their speedy release on a temporary basis into a service deemed essential and experiencing a manpower crisis.

However, there are at least three differences:

1. although the chosen police in 2021 were qualified to drive larger blue light vehicles, they were said to need 24 hours training before being deployed;
2. The discourse of patriotism was no longer used. In 2021 the discourse had become to protect the NHS, so that everyone who needed it could be treated; and
3. Release was potentially by men and women, whereas in 1917 the police was an all-male force.

Furthermore, despite the role of the police having changed, since 1996 it is "to keep the Queen's peace" and "to protect, help and reassure the community" [29], their release in times of crisis to lend appropriately qualified officers continues and was deemed more important than other regular police duties. In the 2021 example, they are seen to protect, help and reassure the community, their assistance entirely relevant according to the London Mayor. They acted in both periods as first responders in an emergency, said to be a legitimate use of the police service. As in

1917/18, their loan into another industry also provided surveillance of the functioning of an essential service under pressure.

The temporary loan of the police to support another industry seen as essential at a time of sudden manpower crisis, is an historical continuity, introduced in Chapter 1 [30]. These comparisons between periods are one way of making sense of the history of an occupational group (see Chapter 1, pp. 2).

References

1 *The Police Review and Parade Gossip* 13 April 1917, p. 119.
2 *Report of the Committee on the Police Service of England, Wales and Scotland.* London: HMSO, Cmd. 253.
3 *The Police Review and Parade Gossip* Parade Gossip 8 December 1916 p. 494.
4 T*he Scotsman* 20 March 1917 p. 6.
5 Lloyd George, D. 1934. *Memoirs* Volume 3. London: Ivor Nicholson and Watson, p. 1269 https://archive.org/details/WarMemoirsOfDavidLloydGeorgeVolIii/page/n263/mode/2up?view=theater (accessed 10/1/2024).
6 Davis, B. J. 2000. *Home Fires Burning: Food, Politics, and Everyday Life in World War 1 Berlin*. University of Carolina Press.
7 Zieman, B 2014. Agrarian Society, Chapter 15 In Winter, J. ed. *The Cambridge History of the First World War,* Volume 2: The State. New York: Cambridge University Press pp. 382–407.
8 Board of Agriculture for Scotland, *Sixth Report of the Board of Agriculture for Scotland for the Year Ending 31st December 1917*, 1918. Cd. 9069.
9 Whetham E. H. 1978. *The agrarian History of England and Wales* Volume VIII 1914–39 p. 44, p. 119–20.
10 Rew, H. 1918. The prospects of the world's food supplies after the war. *Journal of the Royal Statistical Society* 81(1), 41–71 p. 43.
11 Barnett, C. 1970. *Britain and her army, 1509–1970: a military, political and social survey*. London: Allen Lane p. 202.
12 Middleton, T. H. 1923. *Food Production in War.* Oxford: The Clarendon Press, p. 240–2.
13 Ernle, Lord. n.d. *The Land and its People; Chapters in Rural Life and History.* London: Hutchinson, p. 131–2.
14 Dewey P. E. 2014. *British Agriculture in the First World War.* London: Routledge. 2nd edition, p. 100.
15 Abridged from Best, R.H. and Coppock, J.T. (1962) *The changing use of land in Britain*. London: Faber and Faber p. 77.
16 Middleton, 1923, op. cit. p. 255.

17 Whetham, 1978, op. cit. p. 113.
18 Ernle n.d. op. cit. p. 147.
19 Whetham, 1978 op. cit. p. 108.
20 Middleton, op. cit. pp. 228–32.
21 Ernle n.d. op. cit. p. 154–5.
22 Ernle, n.d. op. cit. p. 150.
23 Dewey 2014, op. cit. p. 101–3.
24 Crowe, H. 2008. Keeping the Wheels of the Farm in Motion: Labour Shortages in the Uplands during the Great War. *Rural History*, 19(2) p. 203.
25 Ernle, n.d. op. cit., p. 146.
26 Lloyd George, 1934, op. cit. p. 1322.
27 Neocleous, M. 2014. *War Power, Police Power* Edinburgh University Press. pp. 10–11.
28 *iNews* January 8 2021 https://inews.co.uk/news/uk/covid-police-drafted-in-london-ambulances-uk-suffers-highest-death-toll-823087 (accessed 25/1/2024).
29 The Police Foundation and The Policy Studies Institute, 1996. *The Role and Responsibilities of the Police*. https://www.police-foundation.org.uk/wp-content/uploads/2017/06/roles_and_resp.pdf p. xii. (accessed 25/1/2024).
30 Lawrence, P. 2016. Historical criminology and the explanatory power of the past. *Criminology and Criminal Justice* 19(4): 493–511. https://doi.org/10.1177/1748895818794237 (accessed11/1/2024).

Index

M. Fraser, *The British Police and Home Food Production in the Great War*, https://doi.org/10.1007/978-3-031-58743-6

GPSR Compliance

The European Union's (EU) General Product Safety Regulation (GPSR) is a set of rules that requires consumer products to be safe and our obligations to ensure this.

If you have any concerns about our products, you can contact us on ProductSafety@springernature.com

In case Publisher is established outside the EU, the EU authorized representative is:

Springer Nature Customer Service Center GmbH
Europaplatz 3
69115 Heidelberg, Germany

Batch number: 10755051

Printed by Printforce, the Netherlands